GREAT COMMANDERS OF THE MODERN WORLD

GREAT COMMANDERS OF THE MODERN WORLD

Edited by
Andrew Roberts

Quercus

Dedicated to Alex Coulson and Alec Foster-Brown

First published in Great Britain in 2009 by Quercus

This paperback edition published in 2011 by

Quercus
55 Baker Street
7th Floor, South Block
London
W1U 8EW

A CIP catalogue record for this book is available from the British Library.

ISBN 978 0 85738 591 8

10 9 8 7 6 5 4 3 2 1

Text designed and typeset by Ellipsis Digital Ltd

Printed and bound in Great Britain by Clays Ltd, St Ives plc

CONTENTS

INTRODUCTION TO THE HISTORY OF MODERN MILITARY COMMAND

The twentieth century was the most brutal of all the centuries of human existence, not least because it has been estimated that more people perished violently in that century than in all the previous centuries put together.

Although this fourth and last volume of the *Great Commanders* opens with Helmuth von Moltke's campaigns of the late nineteenth century, these curiously presage the struggles of the twentieth in that his great victory in the Franco-Prussian War of 1870–1 was precisely what Kaiser Wilhelm II hoped to replay in the opening stages of the Great War in 1914. The military weapons of the American Civil War – trenches, railways, barbed wire, above all machine-guns – were to find their apogee in the Great War, as man's natural inventiveness was bent towards destruction. (It was said that Moltke only smiled twice in his life; first when the great Belgian fortress of Liège fell to his forces, and second when he heard his mother-in-law had unexpectedly died.)

An intriguing feature of all four volumes has been how widely the distribution of military genius is spread, both chronologically and geographically. I was surprised that our list contained quite so many non-Europeans, since Europe has been the principal crucible of global warfare for the past half-millennium or so. It was not out of political correctness that warriors such as Shaka Zulu, Tomoyaki Yamashita and Vo Nguyen Giap forced themselves into these pages, but because great military attributes seem to know no racial or geographical boundaries.

This volume of course concentrates on the generals of the two cataclysmic global conflicts that so scarred the twentieth century, and features no fewer than four German generals of the Second World War. This serves to remind us how fortunate we are that despite the generally higher quality of the Wehrmacht in comparison with the Allied generals in the Second World War, the Führer himself was a strategic dunderhead who nonetheless reckoned himself 'the greatest warlord who ever lived'. I have – except in the debatable exception of Eisenhower – opted for hands-on tacticians and battlefield commanders, as opposed to chiefs of staff who decided grand strategy from behind the lines, or even in different countries altogether. Good arguments can be made for General George C. Marshall, General

Sir Alan Brooke (later Lord Alanbrooke) and even for the 'Big Three' statesmen being included as great commanders for the way that they planned the victorious grand strategy of the Second World War at Tehran and Yalta, but they are not generally recognized as such, even though they probably had more overall say in how that war was won than the generals who fought such battles as El Alamein, Stalingrad and Kursk.

So what are the attributes that distinguish history's greatest commanders? 'The power to command has never meant the power to remain mysterious,' wrote one of them, Marshal Foch, in his 1919 work *Precepts and Judgments*. The qualities are not secret but openly on display and have been remarkably unchanging over the centuries, and thus capable of analysis in this book. For all the revolutions in technology, the coming and departing of the Age of Gunpowder, the advent of the machine-gun and the rise of air power, the characteristics have remained astonishingly ageless. Heinz Guderian would have recognized the audacity of Joshua; Dwight Eisenhower would have admired the sheer, overawing puissance of Cyrus; Robert E. Lee would have applauded the attention to detail of Wellington, or for that matter the tactical aptitude of Alexander. Fuller's correspondent – and sometime protagonist – Liddell Hart wrote in his *Thoughts on War* in 1944: 'The two qualities of mental initiative and strong personality, or determination, go a long way towards the power of command in war – they are, indeed, the hallmark of the Great Captains.' In these pages will be found much mental initiative and strong personality, certainly, but also a feel for the coup d'œil, the capacity for inspiring strangers, a remarkable sense of timing, an aptitude for observation, the ability to create surprise, a facility for public relations, the gift of interlocking strategy with tactics and vice versa, a faculty for predicting an opponent's likely behaviour, a capability for retaining the initiative, and as General Patton wrote in October 1944, a capacity for 'telling somebody who thinks he is beaten that he is not beaten'.

It will be noted that no soldier of our own times has merited inclusion in this volume. That is because although the Cold War overheated regionally on a number of occasions – Korea and Vietnam being the most obvious examples – the nuclear deterrent has kept the Great Powers formally at peace, while generalship has tended to become a more managerial, even technological, concept. Nor has the post-1989 world thrown up great commanders of the significance of those featured in this book. In time, the names of General Norman Schwarzkopf in the 1990–1 Gulf War and General David H. Petraeus in the insurgency phase of its successor conflict, the Iraq War, might be seen as justifying inclusion in a future volume, particularly the latter for his command of the 'surge' of five extra US brigades, numbering over 30,000 men, in 2007. At the time of writing in the spring of 2011, with the war in Afghanistan still being fought, it is simply too early to make any such judgement.

Andrew Roberts, May 2011
www.andrew-roberts.net

CONTRIBUTORS

JOHN LEE

John Lee has MAs in War Studies and Social and Economic History. He is an Honorary Research Fellow of the Centre for First World War Studies, University of Birmingham, and a member of the British Commission for Military History, the Western Front Association, the Gallipoli Association, the Army Records Society and the International Churchill Society. He is the author of *A Soldier's Life: General Sir Ian Hamilton 1853–1947* (2001); *The Warlords: Hindenburg and Ludendorff* (2005); and, with his wife, Celia Lee, *Winston and Jack: The Churchill brothers* (2007) together with fifteen or more chapters in books of essays on the First World War. Retired from a life in bookselling and publishing, he is a battlefield guide for Holt's Tours, and a military lecturer and writer.

IAN BECKETT

Ian Beckett is Professor of History at the University of Northampton. Educated at Aylesbury Grammar School and the universities of Lancaster and London, he has been Major General Matthew C. Horner Professor of Military Theory at the US Marine Corps University, Professor of History at the University of Luton and Visiting Professor of Strategy at the US Naval War College, as well as Senior Lecturer in War Studies at the Royal Military Academy, Sandhurst. A Fellow of the Royal Historical Society, he is Chairman of the Army Records Society and Secretary to the Trustees of the Buckinghamshire Military Museum. His many publications on British military history and the First World War include *The Amateur Military Tradition, 1558–1945* (1991); *The Victorians at War* (2003); *Ypres: The First Battle, 1914* (2004); *The Great War 1914–1918* (2007); and *Territorials* (2008). He is currently working on the politics of command in the late Victorian army.

PETER HART

Peter Hart was born in 1955. He went to Liverpool University before joining the Sound Archive at the Imperial War Museum in 1981. His many books include *Jutland, 1916* (2003); *Somme, 1916* (2005); *Bloody April* (2005); *Aces Falling: War above the Trenches, 1918* (2007); and *1918: A Very British Victory* (2008). He is now Oral Historian at the Imperial War Museum Archive.

CHARLES WILLIAMS

Charles Williams, Lord Williams of Elvel CBE, is a Labour peer and political biographer. Educated at Westminster and Christ Church, Oxford, he did National Service in the King's Royal Rifle Corps and went on to a career in banking before becoming Chairman of the Price Commission in 1977. In 1985 he became a life peer and in 1989 was elected Deputy Leader of the Opposition in the House of Lords. His political biographies to date are of de Gaulle, Adenauer and Pétain, and he is currently working on a life of Harold Macmillan. As a former cricketer (he captained Oxford in 1955 and subsequently played for Essex in the County Championship) he thought that he had all the right credentials to write the life of Sir Donald Bradman, a book which Bradman himself said was the best of the many written about him. His wife Jane was Rab Butler's niece and one of Winston Churchill's devoted secretaries. They live in London and Wales.

JEREMY BLACK

Graduating from Queens' College, Cambridge, with a starred first, Jeremy Black studied at St John's College and Merton College, Oxford, before teaching at Durham and Exeter universities. He is a Fellow of the Royal Society for the Arts and a past Council member of the Royal Historical Society. His many books include *War and the World*

1450–2000 (2000); *World War Two: A Military History* (2003); *The British Seaborne Empire* (2004) and, for Quercus, *Tools of War* (2007).

CARLO D'ESTE

Carlo D'Este served in the US Army in Vietnam, Germany and Britain before retiring as a lieutenant-colonel in 1978. He received his MA from the University of Richmond and has now published six books that have won high praise both sides of the Atlantic: *Bitter Victory: The Battle for Sicily, 1943* (1989); *World War II in the Mediterranean 1942–1945* (1990); *Fatal Decision: Anzio and the Battle for Rome* (1991); *Patton: A Genius for War* (1995); *Decision in Normandy: The Real Story of Montgomery and the Allied Campaign* (2000); and *Eisenhower: Allied Supreme Commander* (2003). Acclaimed by Antony Beevor, John Keegan, Victor Davis Hanson, Martin Blumenson and Alan Clark, he is one of the greatest military biographers working today.

ANDREW ROBERTS

Andrew Roberts took a first in modern history from Gonville & Caius College, Cambridge, from where he is an honorary senior scholar and PhD. His biography of Winston Churchill's foreign secretary Lord Halifax, entitled *The Holy Fox*, was published by Weidenfeld & Nicolson in 1991, followed by *Eminent Churchillians* (Weidenfeld & Nicolson, 1994); *Salis-*

bury: Victorian Titan, which won the Wolfson Prize and the James Stern Silver Pen Award (Weidenfeld & Nicolson, 1999); *Napoleon and Wellington* (Weidenfeld & Nicolson, 2002); *Hitler and Churchill: Secrets of Leadership* (Weidenfeld & Nicolson, 2003) and *Waterloo: Napoleon's Last Gamble* (HarperCollins, 2005).

He has also edited a collection of twelve counterfactual essays by historians entitled *What Might Have Been* (Weidenfeld & Nicolson, 2004) as well as *The Correspondence of Benjamin Disraeli and Mrs Sarah Brydges Willyams* (2006). His *A History of the English-Speaking Peoples Since 1900* (Weidenfeld & Nicolson, 2006) won the US Intercollegiate Studies Institute Book Award for 2007. Dr Roberts is a Fellow of the Royal Society of Literature, an honorary Doctor of Humane Letters, and reviews history books for more than a dozen newspapers and periodicals. His *Masters and Commanders: How Churchill, Roosevelt, Alanbrooke and Marshall Won the War in the West, 1941–45* (Allen Lane 2009) won the International Churchill Society Award, and his *The Storm of War: A New History of the Second World War* (Allen Lane, 2010) won the British Army Military Book of the Year Award. His website can be found at www.andrew-roberts.net.

ALLAN MALLINSON

Allan Mallinson was a professional soldier for thirty-five years, first in the infantry and then the cavalry. He commanded the 13th/18th Royal Hussars, whose history, after their 1992 amalgamation with the 15th/19th King's Royal Hussars, he wrote: *Light Dragoons* (published by Leo Cooper); recently republished with an additional chapter covering the years since amalgamation. He is author of the Matthew Hervey series of novels, set in the Duke of Wellington's cavalry, the tenth of which – *Warrior* – was published by Transworld in June 2008. He is currently working on a history of the British Army. Allan Mallinson is a regular reviewer for *The Times*, the *Sunday Telegraph*, and *The Spectator*, and is the defence columnist of the *Daily Telegraph*. He lives on Salisbury Plain.

MICHAEL BURLEIGH

Since 2001 Michael Burleigh has worked as an independent historian and writer after eighteen years as an academic: in Britain at New College, Oxford, LSE and the University of Cardiff, and in the United States at Rutgers University in New Jersey, at Washington & Lee University in Virginia, and as Kratter Visiting Professor at Stanford University, California. His recent books include *The Racial State: Germany 1933–1945* (1991); *Death and Deliverance: Euthanasia in Germany 1900–1945* (1994); *Ethics and Extermination: Reflections on Nazi Genocide* (1997); *The Third Reich: A New History* (2001), which won the Samuel Johnson Prize for Non-Fiction; *Earthly Powers* and *Sacred Causes*, a two-volume study of politics and religion in Europe from the Enlightenment to Al-Qaeda (2005–6); and *Blood and Rage: A Cultural History of Terrorism* (2008). He is a regular commentator on international affairs for *The Times, Sunday Times* and *Daily Telegraph,* and he has made three award-winning television documentaries. He is married and lives in London.

ALISTAIR HORNE

Alistair Horne was educated at Le Rosey, Switzerland, and Jesus College, Cambridge. He ended his war service with the rank of captain in the Coldstream Guards attached to MI5 in the Middle East. From 1952 to 1955 he worked as a foreign correspondent for the *Daily Telegraph.* In 1969 he founded the Alistair Horne research fellowship in modern history, St Antony's College, Oxford. His numerous books on history and politics have been translated into over ten languages; he was awarded the 1962 Hawthornden Prize (for *The Price of Glory*) and the 1977 Wolfson Prize (for *A Savage War of Peace*). In 1992 he was awarded the CBE; in 1993 he received the French Légion d'Honneur for his work on French history and a Litt.D. from Cambridge University. He was knighted in 2003 for services to Franco-British relations.

SIMON SEBAG MONTEFIORE

Simon Sebag Montefiore was in born in 1965 and read history at Gonville and Caius College, Cambridge. *Catherine the Great and Potemkin* (2000) was shortlisted for the Samuel Johnson, Duff Cooper and Marsh Biography Prizes. *Stalin: The Court of the Red Tsar* won the History Book of the Year Prize at the 2004 British Book Awards; its prequel, *Young Stalin*, was awarded the Costa Prize and the Los Angeles Times Book Prize in 2007, and was nominated for the James Tait Black Memorial Prize in 2008. Montefiore's books are worldwide bestsellers, published in thiry-four languages. *101 World Heroes* was published by Quercus in 2007 and has been translated into ten languages; its companion volume, *Monsters*, appeared in September 2008. A Fellow of the Royal Society of Literature, Montefiore lives in London with his wife, the novelist Santa Montefiore, and their two children. His most recent book is *Jerusalem: The Biography*, published in 2011, a fresh history of the Middle East.

RICHARD OVERY

Richard Overy is a best-selling historian. He specializes in the Hitler and Stalin dictatorships, the Second World War, air power in the twentieth century, and German history from 1900. He is the author of *Russia's War* (1997); *Interrogations: Inside the Mind of the Nazi Elite* (2002); *The Dictators: Hitler's Germany and Stalin's Russia* (2004); and *The Third Reich: A Chronicle*. Richard Overy is Professor of History at Exeter University.

TREVOR ROYLE

Trevor Royle is an author and broadcaster specializing in the history of war and empire with over a score of books to his credit. His latest books are *Flowers of the Forest: Scotland and the First World War* (2006); *Civil War: The Wars of the Three Kingdoms, England, Ireland and Scotland, 1638–1660* (2004); and a study of General George S. Patton as a military commander. Other recent books include *Winds of Change: The End of Empire in Africa* (1996); *Crimea: The Great Crimean War 1854–1856* (1999); and a highly praised biography of the controversial Chindit leader Orde Wingate. He is currently writing concise histories of the pre-1968 Scottish infantry regiments. As a journalist he is an Associate Editor of the *Sunday Herald* and is a regular commentator on defence matters and international affairs for the BBC. He is a Fellow of the Royal Society of Edinburgh.

ALAN WARREN

Alan Warren is the author of *Waziristan: the Faqir of Ipi and the Indian Army* (2000); *Singapore 1942: Britain's Greatest Defeat* (2001); and *World War II: A Military History* (2008). He has been a Fellow of the State Library of Victoria and has taught at Monash University, Melbourne, Australia.

HUGO SLIM

Dr Hugo Slim has worked for Save the Children UK and the United Nations in Sudan, Ethiopia, the Middle East and Bangladesh. He has been on the Council of Oxfam GB and an International Advisor to the British Red Cross. He was Reader in International Humanitarianism at Oxford Brookes University from 1994 to 2003 and most recently was Chief Scholar at the Centre for Humanitarian Dialogue in Geneva. He is currently a Director of Corporates for Crisis in London, advising international companies on their community relations in emerging markets. His latest book is *Killing Civilians: Method, Madness and Morality in War* (2007). He is a grandson of Field Marshal the Viscount Slim.

JOHN HUGHES-WILSON

John Hughes-Wilson is one of Britain's leading commentators and authors on intelligence, contemporary military developments and military history. With Professor Richard Holmes, he is President of the Guild of Battlefield Guides and has specialized in a number of First World War topics, from the mystery of Richthofen's death to British undercover intelligence operations. He is a frequent broadcaster for BBC television, and is the presenter of numerous television programmes, including the award-nominated BBC *What If?* television series. His non-fiction books include the best-selling *Military Intelligence Blunders* (1999) and *Blind-*

fold and Alone (2001), described by reviewers as 'the definitive work on the First World War executions'. *The Puppet Masters*, the secret history of intelligence, was published in 2004 to outstanding reviews and was shortlisted for the Westminster Gold Medal for military history. His latest book is *An American Coup: Who Really Killed JFK?* (2008). After initial service as an infantry officer with the Sherwood Foresters, during his twenty-five years in the Intelligence Corps he saw active service in the Falkland Islands, Cyprus, Arabia, and Northern Ireland as well as the political jungles of Whitehall and NATO. His family has grown up and he now lives and works in Cyprus.

MARTIN VAN CREVELD

Martin van Creveld, formerly of the Hebrew University, Jerusalem, is a leading expert on military history and strategy, with a special interest in the future of war. He is the author of twenty books, including *The Culture of War* (2008); *The Changing Face of War: Lessons of Combat from the Marne to Iraq* (2007); *The Transformation of War* (1991); *Command in War* (1985) and *Supplying War* (1978). Between them, these books have been translated into seventeen languages.

He has acted as consultant to defence establishments in several countries, and has taught or lectured at practically every institute of strategic studies from Canada to

New Zealand and from Norway to South Africa. He has also appeared on numerous television and radio programmes as well as writing for, and being interviewed by, hundreds of newspapers and magazines around the world. He is married to Dvora Lewy, a painter, and lives in Mevasseret Zion near Jerusalem.

GEOFFREY PERRET

Geoffrey Perret was educated at Harvard and the University of California at Berkeley and served for three years in the US Army. He is the award-winning author of thirteen books, including *Old Soldiers Never Die: The Life of Douglas MacArthur* (1996); *Eisenhower* (2000); *Ulysses S. Grant: Soldier and President* (1997); and *Commander in Chief* (2007). Three of his works have received Notable Book of the Year awards from the *New York Times* and four have been nominated for the Pulitzer Prize for History. His work has appeared in the *New York Times*, the *Washington Post*, *American Heritage*, *Military History Quarterly*, *Civil War Book Review* and *GQ*. He is currently writing an account of the Vietnam War that focuses on William C. Westmoreland, Nguyen Cao Ky and Vo Nguyen Giap.

HELMUTH VON MOLTKE

1800–91

JOHN LEE

HELMUTH VON MOLTKE, called 'the Elder' to distinguish him from his less distinguished nephew of the same name, brought the Prussian army to a level of machine-like efficiency that saw complete victory in three wars in seven years. In so doing, Moltke played a crucial part in the creation of the modern state of Germany. He also perfected a staff system and military organization that became the model for most armies in most countries of the world today.

Like many a great 'Prussian' general, Helmuth Karl Bernard von Moltke was born outside the kingdom, in Mecklenburg, into a family that had long served in the Prussian army. Because his father owned property in Denmark, Moltke's first military service was in the Danish army. He entered the military academy in Copenhagen in 1811 and graduated fourth in his class in 1819. After duty with two Danish infantry regiments he made a visit to Berlin, where the career prospects in the Prussian army so impressed him that he resigned his commission and applied to join the Prussian service.

The brightest and the best

In 1822 Moltke passed the stiff Prussian army entrance examinations, supervised by the great Gneisenau himself, and became a second lieutenant in the 8th Infantry Regiment. Such was the intensity of his study of the military art that, within a year, he passed into the War School in Berlin, then directed by Carl von Clausewitz. He earned a reputation as a 'library rat', working long and hard at his studies and excelling in modern languages (his mother had been an accomplished linguist). In 1826 he passed his final examinations, and was marked out as a young man of great promise. Rejoining his regiment, he ran a school for cadet officers and impressed all his superiors. From 1828 to 1831 he served in the Topographical Survey directed by the General Staff, doing important detailed work, as well as commencing deep studies of historical campaigns. He also wrote a number of literary essays, and contracted to translate the twelve volumes of Gibbon's *Decline and Fall of the Roman Empire* into German, only to see the publisher fail as he reached volume eleven. Years of work yielded a fee of only £25.

In 1832 Moltke was appointed to the General Staff in Berlin; his promotion to captain in 1834 was four years ahead of his contemporaries. In 1835, on leave of absence to tour Turkey, he was introduced to the chief adviser to the sultan, to whom he explained the idea of war games (*Kriegspiel*) so effectively that the Turkish Porte officially requested his services as a military adviser to their army. For four years he helped to reorganize the army along Prussian lines, conducted survey work and the planning of fortifications, and saw a little active service against the forces of the rebel, Mehmet Ali. In 1839 his sound advice was ignored and he saw the Turkish army defeated at Nisib. He returned to Berlin with his reputation enhanced by the detailed reports he posted.

In 1842 Moltke married Marie Burt, the English step-daughter of his sister Augusta. She was twenty-five years younger than he was and, though childless, they were intensely happy together. The following year he invested all his savings in the new Hamburg–Berlin railway, and wrote an important essay on the principles of running a railway. Moltke was instantly aware of the huge importance this new technology would have, not just on a nation's economy, but on its capacity to mobilize for, and prosecute, war. In 1845 he was briefly the personal aide-de-camp to the aged and dying Prince Henry of Prussia, which brought him more closely to the attention of the royal family.

By 1848 Moltke had served on the staff of VIII Corps and became chief of staff to IV Corps. He had a busy and contented life, keeping his troops at a peak of efficiency, studying, writing and looking forward to retirement. The wave of revolutions that swept Europe that year saw the Prussian army stand firmly behind their king as he resisted the demand for reform from his citizens. In the wake of the 1848 Revolutions, Prussia had to stand by as Austria asserted her dominant role in German affairs, and this humiliation was deeply felt. In 1855 Moltke became the first adjutant to Prince Frederick William, and for two years toured with him through England, Russia and France. In 1857 the prince's father, acting as prince regent for his dying brother, temporarily appointed Moltke as chief of staff to the Prussian army and confirmed him in that post the next year.

Chief of the General Staff

After the catastrophe of 1806 – the defeats by Napoleon at Jena and Auerstädt and the collapse of the Prussian state – a coterie of reforming geniuses comprising Scharnhorst, Gneisenau and Clausewitz came together to lay the basis for a new sense of professionalism in the army. Under this new system the best young officers were singled out for intensive training in staff procedures, and they could thus be depended upon, regardless of which commanders they served, to act in a uniform manner according to sound military principles. The permanent General Staff in Berlin studied all military questions of the day, and rotated their officers through service with the army in the field, so that their methods were spread throughout the army.

Moltke inherited this well-defined system and brought to it a new intensity based on his own studious approach to life and work. Of the 120 candidates a year for the Berlin War Academy, 40 were accepted, and the 12 top graduates were selected for General Staff duties. Even these were only on probation until Moltke personally approved their work. Thus it was not long before he built up a corps of staff officers all completely in tune with his thought and methods. All staff officers were rotated through line units between promotions, and their ethos thus permeated the army. Many brigade and division commanders were trained by Moltke, and every general had a General Staff-trained officer as his chief of staff. It was this uniformity of doctrine and purpose that enabled Moltke to direct the movement of armies larger than Napoleon's with a sure confidence.

The General Staff studied all possible military contingencies that might affect the Prussian state. They drew up war-plans for most eventualities, and planned mobilizations accordingly. It was Moltke's personal view that 'a mistake in the original concentration of the army can hardly be made good in the entire course of the campaign'. His lifelong interest in railway development transformed the capacity of the army to mobilize quickly and effectively for war. After studying the problem in test mobilizations, he decentralized the whole process down to the corps districts into which the army was organized, and made them liaise with local railways to speed up and perfect the process. No new railway was built in Prussia without integrating it into military planning. A special Lines of Communication Department of the General Staff was created to work with a civilian–military Central Commission of Railways, and one of Moltke's three principal assistants was solely responsible for railway timetables and transportation.

Conscription, and the creation of large trained reserves, saw armies swell in size. Together with the increased lethality of firearms, Moltke saw that the ideal of seeking the enemy's flank on the battlefield was no longer possible. Instead he devised the scheme of conducting strategic developments over huge distances by a continuous operational sequence of mobilization, concentration, movement and battle.

Besides war-planning and mobilization, Moltke developed the educational role of the General Staff, through a deep and intense study of military history (one of the earliest special departments of the General Staff) and the use of war games and staff rides to allow officers to develop their skills in appreciating terrain and issuing orders under pressure without being actually involved in fighting.

In 1864 Prussia and Austria waged war on Denmark over the right to rule the provinces of Schleswig and Holstein. Moltke was not greatly involved in the early stages of the campaign, and watched from the sidelines as some reactionary old generals, who openly despised the 'pampered' staff officers, failed miserably to bring the Danes to heel. At this point the king appointed his nephew, Prince Frederick Charles, to command the army, and made Moltke his chief of staff in the field. By dint of careful planning and the proper concentration of troops at the decisive point, the war was brought to a rapid and successful conclusion. The General Staff, having won the confidence of the king, received an important boost to its reputation.

The Seven Weeks War

It was clear that Prussia would soon challenge Austria for supremacy in Germany. The crunch came in 1866, when Otto von Bismarck, the Prussian chief minister, sought a promise of French neutrality and an Italian alliance. Austria saw the threat and began mobilizing her army. The war against 'German brothers', was not popular, and the king delayed the decision to mobilize his army beyond the expected date. Moltke had to adapt his war-plan to cover Prussia from possible Austrian attack, and to deal with

Austria's several allies among the many disunited states of Germany.

Moltke employed the principal railways to deploy three armies towards Austria and her Saxon ally. These armies would advance separately and be brought together in the decisive theatre of operations once it had been ascertained where that was going to be. Though the Austrians had mobilized weeks ahead of Prussia, Moltke was not dismayed: he felt he had a deep understanding of the Austrian army and its commanders, and was confident in the ability of the Prussian army and its generals. In a study of the Franco-Austrian War of 1859 he had written against the name of Benedek, the current Austrian commander: 'No commander-in-chief or strategist: will want a deal of assistance in running an army.'

From Berlin, on 15 June, at the end of lines of telegraphic communication, Moltke directed three armies south, while keeping in touch with three smaller armies dealing with Prussia's enemies within Germany. Even when one of these armies was defeated by the Hanoverians, Moltke did not allow himself to be distracted from his main objective: the defeat of Austria. (The Hanoverians, in fact, had fired off all their ammunition and surrendered two days later.) A complete breakdown of the telegraph left Moltke in a void for a day, and the poor performance of the Prussian cavalry in obtaining information about enemy movements often led to contact being lost with the Austrians.

Moltke took calculated risks based on his solid knowledge of his own army and that of the enemy. He knew that the Austrians were deploying forward into Bohemia in a central mass. He knew that an active and resourceful commander might try to defeat the separate Prussian armies in sequence. But he also knew that his armies were drawing together, so denying the enemy the room to manoeuvre, and he knew that Benedek was indecisive and would need weeks to organize offensive operations.

Despite the slow progress of one army, and the short-term defeat of part of another, the Prussians ground forward relentlessly, inflicting a series of small tactical defeats on the Austrians and destroying what remained of their commander's will to win. Once the Austrians made their stand at Königgrätz, with the River Elbe at their back, the stage was set for victory.

The Battle of Königgrätz

When Prince Frederick Charles's First Army determined to attack the Austrians, he simply asked the Prussian crown prince's Second Army to lend him some assistance on his left. Moltke, now present with the army, immediately ordered the whole of the Second Army to march with all speed to close in on the Austrian flank as the First Army and the Army of the Elbe engaged them in battle. He was entirely clear in his intentions, and his orders were brief and to the point.

Having been lacklustre in advance, Frederick Charles impetuously launched his 75,000 infantry at 180,000 Austrians deployed in strong positions. There were anxious moments on the morning of 3 July as the Austrians made good use of their local superiority. At a critical moment Moltke turned to the king and said, 'Your Majesty will win today not only a battle but a campaign.' By 2 p.m. the Second Army had arrived on the field of battle; by 3.30 all three Prussian armies were advancing; by 6.30 the victory was complete. The gamble of uniting three armies on the field of battle had succeeded.

The Prussians allowed the Austrians to escape. They were not seeking to humiliate their enemy, and brought hostilities to a rapid conclusion, before other European powers, especially France, could intervene.

After his stunning masterstroke against the Austrians, Moltke returned to Berlin, saying: 'I hate all fulsome praise ... I but did my duty.' He now integrated all the armies of the northern German states into the Prussian system, and obliged the armies of southern Germany to conform. Almost at once contingency plans began to be laid for a future war with France, with Moltke and his staff drawing up the most intricate arrangements for mobilization by railway. His aim was to deliver over a million men to the Franco-German frontier.

The Franco-Prussian War

Moltke personally held France accountable for 200 years of war in Europe. War with France was not only inevitable but desirable, he believed. Once again his profound knowledge of the capabilities of all the armies involved led him to plan boldly for complete and early victory: 'Our object,' he said, 'is to seek out the main enemy mass and to attack it wherever found.' Needing to strike before Austria could intervene, he wanted 500,000 troops to be instantly available to meet a maximum of 300,000 French at the outbreak of war. Once again he would direct three armies to move quite separately towards the frontier and use one to pin the French army in place while the other two manoeuvred against the enemy flanks.

Prussia's – and Moltke's – opportunity came in 1870, when a dispute arose over a Spanish request for a Hohenzollern prince – a member of Prussia's ruling dynasty – to ascend their throne. When the French objected strenuously, the Prussian king was willing to let the matter drop. When the French rudely insisted that he promise never to support such an idea again, Bismarck and Moltke conspired to edit his reply to give maximum offence to the Emperor Napoleon III. The French then mobilized their army, placing them in the role of aggressor. At this point the Prussian war-plan swung into action. For two weeks, from 15 to 31 July, Moltke had little to do but read novels as half a million Prussian troops moved into position according to his carefully laid plan.

The French mobilization was chaotic, and their military organization all but collapsed under the strain. They rashly launched an offensive into Germany and briefly occupied Saarbrücken. Moltke quickly had to redraw his plans, and tried to secure an early crushing victory. But here the aggressive response of the Prussian commanders in the field was too prompt. Both First and Third Armies launched frontal assaults on the first enemy they encountered, at Spicheren and Worth respectively. The well-armed French inflicted heavy casualties, but could do nothing to stop the Prussians crowding forward, seeking their flanks and turning them out of strong positions. While nearby French forces waited for orders, all the Prussian units marched to the sound of the guns and fought the enemy where they found them. These two early defeats had a profound effect on French morale and on the minds of their commanders.

Moltke set his armies marching towards the west. He considered whether the French might mass against any one of them but concluded that 'such a vigorous decision is hardly in keeping with the attitude they have shown up till now'. The French were in full retreat and threw themselves into the fortress zone of Metz as a safe haven. Moltke simply bypassed Metz and directed his armies across the French line of retreat. He ordered the Second Army and most of the First Army to close in on the French from the south. On 16 August two corps of the Second Army, thinking they were engaging a rearguard, bumped into the entire French army at Vionville. Only the excellent Prussian artillery, vastly improved since 1866, saved them from serious defeat in the morning. More and more Prussian troops were fed into the battle until it seemed to end in a bloody stalemate at about 7 p.m. The late arrival of still more Prussian formations saw the French driven from the field in disorder.

The next major battle (Gravelotte, 18 August) saw the Prussian First and Second Armies with their backs to Paris facing east, as the French tried to escape westwards. Once again impetuous Prussian generals attacked the enemy too vigorously, before the whole army had deployed for a united effort. After hard fighting and grievous losses, they could claim the terrible day as a victory. The French army retreated into Metz and was besieged there.

Moltke then directed fresh forces steadily westwards until contact was made with a new French army supposedly trying to relieve Metz. By 31 August the French had gone into camp on the Meuse at Sedan. There they were completely surrounded and bombarded into submission. On 1 September 104,000 French troops – along with their emperor – passed into captivity. The road to Paris was open.

Paris was duly besieged. Across France the government of the recently declared Third Republic organized new armies, but these were defeated one after another. Metz surrendered and, after an armistice in January 1871, the war came to end in March.

Moltke's legacy

The German empire was declared at Versailles in January 1871. Moltke would continue to serve as chief of the General Staff until 1888, having had a request to retire in 1880 refused. Constantly aware that the French might seek to revenge their defeat at any moment, he planned for every eventuality of war with France and any combination of her allies. Conscious of the dangers of a war on two fronts, he perfected his doctrine of the strategic offensive (carrying the war into the enemy's own territory) and the tactical defensive (using the power of the defence to break the enemy's attacking army before delivering a counter-attack at the decisive moment).

Moltke's legacy is that of the perfect professional, so deeply imbued with knowledge of military history and modern technology that, while never having personally commanded troops in battle, he could execute his plans with an icy calmness, with an intellect that coped with any emergency, and with a clarity of purpose that carried him to victory again and again. The 'Great Silent One', as the German people called him, died in 1891.

GARNET WOLSELEY

1833–1913

IAN BECKETT

WHEN GEORGE GROSSMITH performed the role of the Major-General in Gilbert and Sullivan's The Pirates of Penzance *for the first time at the Opéra Comique on 3 April 1880 no member of the audience would have doubted for a moment that the 'very model of a modern Major-General' was meant to be Garnet Wolseley. Already known to the Victorian public as 'Our Only General', Wolseley had won his reputation on the Red River expedition in the Canadian northwest in 1870, the Ashanti (Asante) War of 1873–4 and, most recently, in bringing the Zulu War to a successful conclusion in 1879. Indeed, Wolseley remained in South Africa overseeing the post-war settlement until returning to England to become quartermaster general at the War Office in July 1880. Yet, his greatest military triumph was yet to come for, in September 1882, his victory at Tel-el-Kebir secured British control over Egypt and the Suez Canal.*

Subsequently, there was also to be failure, for two years later, when the Mahdist revolt broke out in the Sudan, Wolseley was dispatched belatedly to relieve Charles Gordon, besieged in Khartoum: his troops reached the city just two days too late. Unfortunately, Wolseley's health was declining by the time he became commander-in-chief of the British army in 1895, and he had inherited a post stripped of its former powers. Rapid mobilization at the start of the South African War in 1899 owed much to Wolseley, even if subsequent early defeats illustrated how much more military reform was yet required. Nonetheless, he had done much to define national military policy, and laid the foundations for further professionalism.

Born in Dublin on 4 June 1833, Garnet Joseph Wolseley was the eldest of seven children. His father, an impoverished major in the 25th Foot, had left the army shortly after his marriage and died when Wolseley was just 7 years old. The family was left in straitened circumstances, and Wolseley's first commission as ensign in the 12th Foot on 12 March 1852 was secured through his mother seeking a direct nomination from the commander-in-chief, the Duke of Wellington, on the strength of his father's service. At a time when many commissions were still purchased,

Wolseley was to advance without benefit of money through sheer courage and determination, a conviction that his destiny was willed by God having been imbued in the young Wolseley by his mother's deeply held Irish Protestant religious faith. In the Second Burma War (1852–3) Wolseley received a thigh wound that troubled him all his life. In the Crimean War (1854–6) he lost the sight of his left eye. After further service in the Indian Mutiny (1857–8) and the Third China War (1860–61), Wolseley emerged a brevet lieutenant colonel after just eight years service.

The 'Wolseley Ring'

Stationed in Canada, Wolseley visited the Confederacy in 1862, writing memorably of his impressions of leading Confederate commanders such as Robert E. Lee and 'Stonewall' Jackson. He also consolidated his reputation as a military reformer in 1869 with a practical manual, *The Soldier's Pocket Book,* aimed at improving tactical efficiency. The manual demonstrated Wolseley's instinctive understanding of logistics, knowledge immediately required when he was chosen to quell a rebellion by French-speaking *métis* in the Canadian northwest in 1870. Wolseley forged through 600 miles of wilderness with a force of 1,200 regulars and militia to re-occupy Fort Garry on the Red River and to return before the lakes froze. As it happened, the *métis* fled before Wolseley reached Fort Garry, but it had been a triumph of minutely supervised organization. Wolseley received a knighthood and was brought back to the War Office as assistant adjutant general in May 1871.

The reforms of the army instituted by Edward Cardwell as secretary of state for war were mostly completed before Wolseley returned to Britain, but he fully supported their aims, such as the introduction of short service. He was then selected by Cardwell in August 1873 to lead the expedition to repel an Ashanti incursion into the Gold Coast, and to punish the Ashanti by advancing on their capital at Kumasi. Once more it was a question of overcoming climate and terrain: Wolseley had to get British troops to Kumasi and back before the onset of the rains and before tropical diseases took their toll. In Canada Wolseley had begun to make note of able young officers such as Redvers Buller and William Butler, and he now summoned them to join him on the Gold Coast. This was effectively the beginning of what was to become known, with the addition of others such as Evelyn Wood, George Colley, Frederick Maurice and Henry Brackenbury, as the 'Wolseley Ring' (or the 'Ashanti Ring').

Wolseley did not always have a free choice of staff on his campaigns, but he preferred those familiar with his working methods and in whom he had confidence. Ultimately, he did become something of a prisoner of the initial success of the 'Ring' system, often employing the same men in case it was felt his rejection of them would reflect on his initial choices. Moreover, Wolseley's ability to manage a campaign decreased in proportion to the increasing scale of the operations with which he was tasked, with many of his chosen subordinates proving unable to act on their own initiative. To rivals, the 'Ring' was the 'Mutual Admiration Society', but it was one of several such networks within the army, reflecting the absence of a properly formed general staff.

The outcome of the Ashanti War was taken to be proof of the success of the Cardwell reforms, although the regulars involved were soldiers enlisted under the old long-service system. Certainly, it was a model campaign and one extensively reported in the press. Wolseley arriving at the Gold Coast in October 1873, using locally raised forces to push the Ashanti back across the River Pra, and making extensive preparations for the arrival of his British troops. Crossing the Pra in January, Wolseley fought two sharp actions, took and burned Kumasi, and was back on the coast by 19 February 1874. Among other rewards, Wolseley was promoted to major-general.

The War Office and reform

In 1875 Wolseley was sent to administer Natal, and then in 1878 to occupy Cyprus, awarded to Britain by the Treaty of Berlin as a potential base should the Russians attempt to control the Dardanelles. Had a major war broken out against Russia, Wolseley would have been chief of staff to the proposed British expeditionary force. In June 1879 Wolseley and his adherents were dispatched to retrieve the situation in Zululand following early British defeats, though the Zulus were defeated before Wolseley reached the front. Completing the subjugation of the Zulus, Wolseley became successively quartermaster general at the War Office in July 1880 and adjutant general in March 1882, despite the opposition of military conservatives, who resented Wolseley's public championship of reform and their exclusion from his now well-publicized campaigns. Indeed, he had earned the particular enmity of the army's long-serving commander-in-chief (and cousin to Queen Victoria), the Duke of Cambridge. The duke resented Wolseley's flouting of command and staff selection by seniority, and believed that long service was essential to discipline, that discipline and drill were the key to military efficiency, and that reform would generally undermine regimental tradition. Wolseley had hopes for the chief command in India, but the duke steadfastly resisted this, while the queen also opposed granting Wolseley a peerage.

In his two successive appointments at the War Office between 1880 and 1890, therefore, Wolseley found it difficult to achieve progress. He did, however, oversee the modernization of the infantry drill book, the introduction of mounted infantry, better tactical training and improvements to the suitability of campaign dress. The most significant development was the extension of the intelligence department of the War Office and the preparation of proper mobilization plans for the home army. Wolseley, indeed, embraced what became known as the 'imperial school' of British strategic thought, placing the priority on home defence and envisaging any potential war against Russia as being waged primarily through amphibious operations. By contrast, the 'Indian school', most often represented by Wolseley's most prominent military rival, Frederick Roberts, feared the Russian threat to India and, therefore, saw operations on the Northwest Frontier as the empire's first priority in war. In the process of his advocacy of proper military planning, Wolseley secured a definitive statement of the purposes for which the army existed and of the relative priorities to be accorded them through the so-called Stanhope Memorandum of 1888.

Egypt and the Sudan

Wolseley's tenure as adjutant general was twice interrupted by further field command. The growing power of Ahmed Arabi and anti-European riots in Alexandria in June 1882 persuaded Gladstone's government that the security of the Suez Canal and European investment in Egypt was imperilled. The British and French fleets, lying off Alexandria in a demonstration of force, were then threatened by the construction of new Egyptian shore batteries. Exceeding his instructions, Admiral Seymour bombarded the batteries on 11 July. With the khedive taking refuge on Seymour's ships, Arabi seized power.

Wolseley, given command on 4 August, had already drawn up a plan for securing the Suez Canal and Ismailia, and advancing on Cairo from the east, using the Ismailia-to-Cairo railway and the Sweetwater Canal to build up supplies before the final advance. The extensive logistic preparations began at once, including the provision of locomotives and wagons to bypass any need to rely on Egyptian rolling stock, and the purchase of horses and mules. Wolseley did not get all his own way on appointments, since the principal commands were taken by those already designated to act in the forthcoming autumn manoeuvres. He also found it politic to accept the queen's son, the Duke of Connaught, as a brigade commander.

Before his arrival at Alexandria, Wolseley, fearing that the Egyptians might block the Suez Canal, ordered a series of feints by the troops that had already landed, in order to persuade Arabi that he would advance upon Cairo from the west. On 19 August the field force embarked, seemingly to land at Aboukir in the west but, under cover of darkness, instead steamed east to Port Said, where marines and seamen had occupied key points. Having secured the canal, Wolseley advanced towards Kassassin to clear the line of the railway and the Sweetwater Canal on 24 August. Arabi entrenched at Tel-el-Kebir, an Egyptian sortie on 28 August being disrupted by a celebrated 'moonlit' charge by the Household Cavalry. A further Egyptian advance was also pushed back on 8 September.

Wolseley had some 16,000 men, while Arabi had perhaps 20,000 men and 75 guns behind entrenchments. Accordingly, Wolseley resolved upon a daring night march across the desert. Having carefully estimated the distance to be covered before dawn, Wolseley started his men out at 1.30 a.m. on 13 September in strictly enforced silence and guided by a naval officer using the stars as reference. The Highland Brigade was spotted coming up to the Egyptian trenches at about 4.55 a.m. In places there was fierce resistance to the Highlanders' charge, but by 6 a.m. the Egyptians were in flight. Wolseley unleashed his cavalry, which entered Cairo on 14 September, Arabi surrendering that night. The British suffered only fifty-seven killed, forty-five of them Highlanders. He received promotion to full general and, at last, the coveted peerage as Baron Wolseley of Cairo and Wolseley.

Though Britain controlled Egypt, the restored khedive still nominally governed. The khedival government had been struggling against the Mahdi's rebellion in the Sudan even before Wolseley's victory at Tel-el-Kebir. Failure to contain the revolt, not least the destruction of an Egyptian army commanded by a former British officer, William Hicks, in November 1883, led the British government to resolve to abandon the Sudan. Sent to evacuate the Egyptian garrisons in January 1884, Charles Gordon chose to hold Khartoum instead. Wolseley, who had seen Gordon off on his mission, urged a relief expedition. It was not until September, however, that he was appointed to command such an expedition.

Influenced by the Red River experience, Wolseley chose to advance down the Nile from the Egyptian frontier rather than crossing the desert from Suakin on the Red Sea coast to the Nile at Berber, as most advised. Indeed, he brought over Canadian *voyageurs* (boatmen) to handle the 800 special boats he intended to construct. This time, however, the logistic problems were even more formidable than in his previous campaigns, since there were 1,600 miles between Cairo and Khartoum. Refused permission by the government to go as far forward as he felt necessary, Wolseley had to rely more upon his chosen subordinates, now grown in seniority and status and riven by mutual animosities. Precious days were lost, especially when the able Herbert Stewart, commanding the 'river column', was mortally wounded. Stewart's successor reached Khartoum on 28 January 1885, two days too late to save Gordon, a particular hero for Wolseley, for whom Gordon's death was a bitter blow. It was little consolation that he was elevated to a viscountcy in August 1885.

Commander-in-chief

Approaching the end of his appointment as adjutant general, Wolseley was finally being considered for command in India – but, as there seemed less likelihood now of war with Russia and as Wolseley's daughter was coming out that season, he declined. Instead, he took up the Irish command in October 1890 and, with increased leisure time began to write more, penning a book on Napoleon and completing two volumes of what was to be an uncompleted life of Marlborough. He was promoted to field marshal in May 1894.

Ambition, however, remained, and it was clear that Cambridge would soon retire as commander-in-chief. Ironically, it was now Buller, his old colleague in the 'Ring', rather than Roberts who was Wolseley's greatest rival for the appointment, especially as the Liberal government favoured Buller. Wolseley regarded Buller's apparent willingness to accept the appointment as something of a betrayal. In the event, a fortuitous change of government saw Wolseley secure the prize, and he became commander-in-chief in November 1895.

Wolseley's period as commander-in-chief, however,

was overshadowed. Firstly, five years previously the Hartington Commission had recommended abolition of the post altogether and its replacement by a Continental-style chief of the general staff. The proposal had died with the opposition of the Duke of Cambridge and the queen, but the powers of the commander-in-chief had been reduced earlier in 1895 so that other departmental heads within the War Office would now sit on an army board with the commander-in-chief and have equal access to the secretary of state for war. Wolseley had no success in trying to restore the supreme authority of his post, describing the new situation as being like the 'fifth wheel on a coach'. He also enjoyed a poor relationship with the secretary of state, Lord Lansdowne, who was a friend and supporter of Roberts. Secondly, Wolseley's health began to fail: a serious illness in 1897, from which he never fully recovered, impaired his memory.

Nevertheless, Wolseley was able to take pride in the efficiency and speed with which the army was mobilized for war with the Boers in October 1899, despite the government's failure to heed his advice to mobilize much earlier. Yet the lack of empathy with Lansdowne had impaired strategic decision-making, and Wolseley's achievement was soon undermined by the early defeats suffered by Buller's field army in the 'Black Week' of December 1899. As a result, Wolseley was not consulted when Roberts was appointed to supersede Buller in South Africa. Wolseley was persuaded to remain at his post by the queen, finally retiring in November 1900 and, to his chagrin, being succeeded by his rival, Roberts. He died at Menton in France on 20 March 1913.

Assessment

Wolseley was a complex character, respected rather than liked by his subordinates. He was not as radical as opponents believed, and was certainly no liberal. Fiercely patriotic, he was utterly contemptuous of politicians and made little secret of his belief that party politics was 'the curse of modern England'. Some saw him, indeed, as a man of Caesarist tendencies, but, in reality, he fully understood the restricted constitutional parameters in which the army existed. That did not prevent him from manipulating the press he affected to despise and playing politics himself through that manipulation. He could be charming when he wished, but could also appear egotistical and snobbish.

Though Wolseley's reputation was to be eclipsed through the subsequent prominence of Roberts and the way in which memories of colonial warfare were forgotten amid the greater impact of the First World War, he did lay important foundations for a more professional army. Sadly, however, it was an unfulfilled career for the leading soldier of his generation. Through no fault of his own, he never exercised the supreme test of command against an equal adversary in the field.

ERICH LUDENDORFF

1865–1937

JOHN LEE

ERICH LUDENDORFF, inseparably linked to the name of Field Marshal Paul von Hindenburg, achieved fame by a string of victories on the Eastern Front between 1914 and 1916. He then brought his formidable energy to the task of organizing Germany for total war and, in 1918, carried his nation to the very brink of victory, before leading her to catastrophic defeat.

The great strength of the Prussian military system was that it harnessed the brightest and the best of its non-noble citizens into an elite dominated by the Junkers – the aristocratic class that had served their kings loyally and well in a unique social contract since the seventeenth century.

The consummate staff officer

Erich Ludendorff, the third of six children, was born on 9 April 1865 at Kruschevnia in the Prussian province of Posen (now part of Poland). His father was descended from Pomeranian merchants and was a landowner of small means, holding a commission in the reserve cavalry; his mother was from an impoverished noble family. He was described as a lonely boy at school, where his obsessive cleanliness separated him from the other boys, and where he displayed a marked talent for mathematics. His father greatly approved of his decision to embark on a military career. In 1877, aged 12, he passed the examinations for a cadet school with distinction, thanks to his excellent mathematics paper, and was placed in a class two years ahead of his actual age. Foregoing the temptations of the sports field and gymnasium, he studied hard and performed well. In 1880, still only 15, he entered the military academy at Lichterfelde, near Berlin, and forged ahead academically. His devotion to study and hard work intensified, driven by iron self-discipline, and he was consistently first in his class. In 1885 he was commissioned into the 57th Infantry Regiment at Wesel.

Eight years of regimental duty followed, first in the infantry, then in the 2nd Marine Battalion, and finally with the 8th Grenadier Guards. Ludendorff was singled out for praise by all his commanding officers. When he was selected for the War Academy in 1893 he flourished in its intense atmosphere. Though marked down for service on the General Staff, he was first posted as a captain commanding a company of infantry at Thorn in 1895, then to other troop staff postings, with 9th Division at Glogau and V Corps at Posen. He was a major by 1900, and served at V Corps headquarters as a senior staff officer from 1902 to 1904.

In 1905 Ludendorff was called by the new Chief of the General Staff, Count Alfred von Schlieffen, to serve in the Second Section, responsible for the mobilization and concentration of all the German armies in the event of war. He loved this painstaking, meticulous work, and by 1908 he was appointed head of the section, and in this role was intimately involved in perfecting the details of what would become known as the Schlieffen Plan. There were many variations of Germany's war-plans, to cover as many eventualities as could be envisaged. The basic problem was how to face war on two fronts, against France in the west and her Russian ally in the east. Calculating that the Russian mobilization would be much slower, the Germans gambled on covering East Prussia with relatively light forces and throwing the vast majority of their army against France, to force a speedy decision on that front, before switching back to the east to settle matters there. They placed their faith in the perfection of their rapid mobilization and deployment of an overwhelming force before their enemy would be fully ready. It was Ludendorff's railway timetables that delivered these troops.

Ludendorff was a full colonel by 1911, acting well above his rank, and so completely in the confidence of Schlieffen's successor, Helmuth von Moltke 'the Younger', that he was widely tipped to become his chief of operations once the 'inevitable' war broke out. Instead he got involved in the political lobbying of the Reichstag for an increase in the size of the army and, in January 1913, was punished by being removed from the General Staff and put in command of an infantry regiment. By April 1914 his natural talent saw him commanding a brigade, with the rank of major-general.

Under fire

On the outbreak of war in August 1914 Ludendorff was, because of his intimate knowledge of the unfolding of the war-plan, posted to Bülow's Second Army, to oversee the vital capture of the Belgian forts at Liège. Though having never been in action before, he performed remarkably well. He led troops forward under fire and personally penetrated to the city's citadel and demanded the surrender of the astonished garrison.

Soon afterwards, news came of Russian successes and some German panic in East Prussia. The high command there was to be replaced and Ludendorff was appointed the new chief of staff to the Eighth Army. On the train to the east he collected his new commander, Paul von Hindenburg. Such was the uniformity of staff thinking in the German army that, by consulting maps at a distance, Ludendorff came up with a plan to save the day which was almost exactly the same as that already being embarked upon by the more vigorous members of the Eighth Army's staff. The result was the Battle of Tannenberg, fought between 23 August and 2 September 1914, which resulted in the annihilation of the Russian Second Army, and the rapid retreat of the First.

Great objectives often require great risks. With calculated boldness based on confidence in their own troops and railway engineers, and an understanding of the strengths and weaknesses of their enemy, the Germans left a very thin screen of cavalry and Landwehr reservists to watch the slowest of the Russian armies (the Russian First Army under Paul von Rennenkampf). The bulk of their forces were swung south, by rail and road, to descend upon the unfortunate Samsonov's Second Army.

Local East Prussian troops, defending their homes, soaked up the Russian offensive while German formations moved into place on each flank. The German field commanders, used to being allowed to conduct their own operations within the wider plan, frequently disobeyed orders from above, but usually to good advantage, as they knew the local conditions. Ludendorff, occasionally prey to bouts of nervous tension as he awaited developments, gave his field commanders freedom of action so long as their fighting contributed to the success of the overall plan. The Russian Second Army was driven to complete destruction in the dark forests of East Prussia by 30 August.

Having received reinforcements from the west, Ludendorff immediately reversed the process and swung the German troops north behind the Masurian Lakes to destroy Rennenkampf's First Army. By 14 September only the speed of the Russian retreat saved them from Samsonov's fate. Delivering the historic province of East Prussia from 'the Cossacks' made Hindenburg and Ludendorff heroes throughout Germany, at a time when heroes were badly needed.

After his victory at the Battle of Tannenberg, the skilled use of the railway system to move troops between threatened points enabled Ludendorff to defeat further Russian offensives in Poland. As the commanders in the east clamoured for more men and matériel to complete their victories, they clashed repeatedly with the new chief of the General Staff, Erich von Falkenhayn. When he did release reinforcements for the Eastern Front Falkenhayn directed them to commanders other than Hindenburg and Ludendorff. Ludendorff argued, with some validity, that these attacks merely drove the Russians back into their limitless country, whereas he planned great turning movements aimed at surrounding and destroying their armies in the field.

This bickering at the highest levels went on into 1916. By then Falkenhayn's reputation had been damaged by his lack of success at Verdun and his failure to allow for Romania's entry into the war. When the Kaiser asked Hindenburg and Ludendorff for their advice, Falkenhayn resigned.

The call to supreme command

On 29 August 1916 Hindenburg became Chief of the General Staff, and Ludendorff, as first quartermaster general, became Hindenburg's deputy and effectively his executive chief of operations. Having been completely absorbed in the complexities of war on the Eastern Front, the two men now had to view the struggle in its entirety. Ludendorff toured the army commands in the west and quickly realized the enormity of the problem that Germany faced. While the battles raged all summer along the Somme, he authorized the construction of massive defences in the rear (the Hindenburg Line), to which he authorized a retreat early in 1917. Proud of his claim to be an infantry officer above all things, he had a special interest in the development of the tactics of the attack and the defence. He oversaw the learning of lessons from the recent fighting, and the drawing up of new training doctrines for a more flexible defence, aimed at first disorganizing the attackers and then defeating them with heavy counter-attacks. This was achieved by inaugurating a debate amongst the staff and combat officers of the armies in the west, and distilling their practical and theoretical wisdom into new instruction manuals for conducting the defence in depth. The old Prussian tradition of studying operations in great detail, to draw lessons and improvements for future practice, was put to good use.

The next great task was the mobilization of the whole nation for war. The General Staff became intricately involved in the organization of the war industries, manpower, the press and, increasingly, domestic politics. Ludendorff's insatiable appetite for work drew one colleague to say that 'Too much rests on Ludendorff: all domestic and foreign policy, economic questions, the matter of food supply, etc.' The army High Command encouraged the imperial navy to begin unrestricted submarine warfare, despite the danger of bringing the United States into the war against them. When the chancellor, Theobald von Bethmann-Hollweg, protested, they engineered his dismissal and had him replaced by a nonentity, more subservient to their will. The German military increasingly intervened in domestic politics to insist on its own viewpoint, and several more leading politicians fell victim to this drive for a united war effort.

1917 was a year to restore German confidence in victory. Their forces so completely defeated a French offensive on the Aisne that the French army became mutinous. The cumulative strain of defeat on the Eastern Front forced Russia into revolution and peace negotiations. Ludendorff's skilful use of the railway to move his reserves about saw a surprise offensive at Caporetto drive Italy to the brink of catastrophe (October–November). The abiding problem was the relentless offensive power of the British Expeditionary Force (BEF) under Sir Douglas Haig. Though Germany managed to contain British attacks at Arras in April, Messines in June, Ypres for three and a half months that autumn and Cambrai in November, the appalling drain on German resources was intolerable. Despite finessing their defensive tactics, Ludendorff could do little to stop the violently destructive nature of the British assaults. The submarine war failed to cripple the British war effort, and the Americans entered the war against Germany in April 1917. The longer the war lasted, the stronger the Allies would become, and the weaker the German home front, suffering terrible deprivation because of the Allied blockade.

The final gamble

Ludendorff determined to force a decision in the west, bringing the Allies to the peace table from a position of strength, before the American army could deploy its full capacity. Able to draw troops away from the Eastern Front, he planned a series of powerful offensives aimed at defeating his most implacable foe, the British. His most remarkable achievement was to reconfigure completely the western German armies for offensive warfare after years of defensive fighting. Having been impressed with the specially trained 'storm troop' battalions in service on the Western Front, with their skills at infiltrating enemy positions to great depth, he arranged for their techniques to be taught to the divisions selected for the upcoming offensive. In the attack divisions, men over 35 were replaced by younger soldiers, and Germany's best weapons and equipment were provided. In particular, Ludendorff selected for army command those generals who had shown their skill at conducting counter-attacks or offensives in France, Russia and Italy. Finally, he brought in Germany's greatest artilleryman, Georg Brüchmuller, who had perfected devastating bombardment programmes that had produced spectacular results in the east.

Although this was to be a final attempt to force the Allies to the peace table, Ludendorff was not able to bring as many new troops to the west as he might have liked.

The High Command had so completely taken over the function of the civilian government, running the economy, social organization and foreign policy, that it was distracted by the possibilities of building on its successes in the east. Many German troops were heavily embroiled in fighting in Finland, the Baltic states and the Ukraine, when everything should have been concentrated for the great effort in the west.

When the attack, Operation Michael, was launched on 21 March 1918 the immediate plan was to drive a wedge between the British and French armies, and turn north and drive the British back through their base camps to the Channel coast and 'into the sea'. Since this was their last chance, the Germans concentrated more guns, attack divisions and aircraft than had ever been used in one sustained attack on the Western Front. The intricately designed bombardment was of an unparalleled intensity, with heavy reliance on gas shells to neutralize the defences without ploughing up the ground with high explosive. Striking the British Fifth Army, the most thinly spread and least prepared in the BEF, the Germans gained more ground in three or four days than any offensive in the west to date. The problem was that, while capturing ground and prisoners did much for German morale, the BEF did not break decisively, and the French rushed to its aid and maintained contact throughout. The second phase of the offensive, Operation

Mars, launched on 28 March, was meant to crack open the British lines at Arras and start the process of rolling up their defences from south to north. It was a massive tactical failure and, soon afterwards, Operation Mars was wound down.

Another powerful attack was launched across the Lys on 9 April, finding a weak spot manned by Portuguese troops. The fighting was so desperate that the imperturbable Haig issued a 'Backs to the Wall' message, calling on his troops to fight to the last man and the last bullet. As this attack ran out of steam, Ludendorff rethought his plan and, in an effort to draw reserves away from the British front, launched a series of attacks against the French. When these threatened Paris itself, it looked as if some great and decisive result must soon be his.

But the effort had worn out the German armies. Ludendorff had let each attack run its course, without concentrating on the original aim of turning the flank of the BEF and driving it to the north. The dogged resistance of the BEF and the nurturing of the French back to an offensive capability, together with the arrival of large numbers of American troops, had frustrated the great hopes of the 'peace offensives' of the spring of 1918.

Defeat and after

Despite the enormous conquests in the east, and advances in the west unheard of since 1914, the Germans now found themselves assailed by powerful and resurgent Allied armies on the Western Front. The proven skill of the German soldiery in defence was not enough to prevent the relentless Allied advance towards the German frontier. Ludendorff's great effort to mobilize the whole nation for total war, to bring the Allies to the peace table, had failed. He suffered greatly from the strain and was ordered by doctors to give up much of his astonishing workload. He offered his resignation once too often to the Kaiser, who accepted it on 26 October. As Germany slid towards defeat and revolution at home, Ludendorff quietly slipped out of the country.

From Sweden he wrote his war memoirs, dedicated to 'The Heroes who Fell Believing in Germany's Greatness', justifying his efforts to galvanize the nation to ever greater efforts to win the war. On his return to Germany he flirted with the new right-wing political movement led by Adolf Hitler. Although he went on to serve as a National Socialist deputy in the Reichstag, he had distanced himself from the Nazis by the time of his death in 1937.

If his life's work ended in catastrophic failure, Ludendorff deserves to be assessed as the product of the most efficient military General Staff the world had seen to date. At every stage in the Great War, Germany, facing a sea of enemies and linked to relatively weak allies, was hugely outnumbered. She defended large gains in France and

Belgium for four years, conquered Serbia and Romania, hammered the Russian armies into a state where socialist agitators were able to seize power and take Russia out of the war, inflicted meaningful defeats on France and Italy, and launched an offensive in 1918 that overshadowed anything that had gone before. The trade blockade imposed by the Allies brought ruin to German agriculture, and her mighty industry had to perform miracles of improvisation to maintain the war effort. But shortages turned into real starvation, and the General Staff found itself doing the job of Germany's weak parliamentary democracy, running the country while trying to run the war. It was too much, even for a man with Ludendorff's Herculean appetite for unremitting hard work.

FERDINAND FOCH

1851–1929

PETER HART

FERDINAND FOCH was a famed academic tactician who had the mental strength cheerfully to abandon his own theories when they proved inappropriate to real life. Although often caricatured as an archetypal individualistic Frenchman, Foch proved a brilliantly successful Allied supreme commander able to form and maintain solid working relationships with his various subordinates, to overcome petty nationalistic disputes, and to drive them on to victory in the final series of battles that secured victory for the Allies in 1918.

Ferdinand Foch was born on 2 October 1851, and so was already 62 when the Great War began in 1914. His formative influences were a splendid education, which fed his gimlet mind; a staunch traditional Catholicism, which gave him a willingness to believe the impossible; and the trauma of French defeat in the Franco-Prussian War of 1870–71, which cemented his deep patriotism. The whole of his military life was devoted to countering the threat from Germany and in seeking revenge for the humiliating loss of Alsace-Lorraine. Foch did not pursue a career serving in the French colonies, as did many of his contemporaries. Minor colonial wars were a mere distraction from the real challenge of defeating Germany.

Theoretical master

In 1885 Foch attended the staff course at the prestigious École Supérieure de Guerre in Paris. His studies convinced him that the only way to counter the perceived German superiority in numbers and matériel was to establish a culture of mental superiority in the French soldier that could overcome physically superior forces by dint of sheer élan in battle. After a tour of duty, Foch returned to the École Supérieure de Guerre, where he was an instructor from 1895 to 1901. Here his vibrant lectures created a sensation amongst a whole generation of French staff officers. Unfortunately, alongside much valuable material on the principles of war, Foch included elements that were demonstrably nonsense. As he had no practical combat experience, he failed to appreciate the coruscating power of modern weapons. He believed in the 'superiority' of the

French national character, the power of 'faith' and the 'supremacy' of the offensive in its ability to seize the initiative and thereby dominate a passive enemy, who could do nothing but endure. His arguments were difficult to counter without suggesting, for instance, that the French character was not superior – a cleft stick for even the most recalcitrant of his students. In the event few dissented. Foch offered them hope, a way of succeeding against the hated Germans.

Foch's military career stalled owing to his staunch Catholicism, and also because of widespread suspicions that he was anti-republican. However, he managed to impress the secular, indeed downright anti-clerical, prime minister, Georges Clemenceau, who approved his appointment in 1907 as commandant of the École Supérieure de Guerre, a post he held until 1911. Now Foch's influence began to really permeate the French High Command. This was to have unfortunate consequences. The desire for a more aggressive approach to the conduct of war led them into dangerous waters with the emergence of 'Plan XVII', which envisioned a ferocious French assault straight into the 'lost' province of Lorraine. This was certainly aggressive enough, but it offered far too many hostages to fortune, by ignoring many of the more sensible principles of war hammered out by Foch in his lectures. As a result, Foch himself harboured reservations over the plan, in particular the assumption that the Germans would respect Belgian neutrality, which thereby exposed the French to the threat of a comprehensive strategic surprise if the Germans attacked through Flanders. Foch was also worried that the French would not have sufficient numerical supremacy to make success likely against their well-armed German counterparts.

The real thing: August–September 1914

When the war began in August 1914, Foch was the commander of XX Corps, based at Nancy. He was in his element at last, for despite his doubts over Plan XVII he was more than ready to launch the assault – finally to test himself and his theories on the field of battle. Foch was as caught up in the emotion of the moment as anyone else and, although he had recognized the risk of the Germans attacking through the Low Countries, he failed to think it through. As the French army lunged to the east they fell straight into the German trap. The Germans wheeled through Belgium into northern France and headed straight for Paris. Foch's part in the French offensive was initially successful, but the awakening was to be brutal. XX Corps became isolated, and when the Germans launched a counter-attack on 20 August, the result was a disaster. For a while, Foch's whole reputation and career as a general hung in the balance. Had it not been for the wholesale cull of incompetent French generals and the personal backing of the commander-in-chief, Joseph Joffre, he might well have lost his command.

Thanks to the enduring confidence of Joffre, Foch was appointed to command a scattered mélange of units that would eventually form the Ninth Army in the area south of the River Marne, not far to the east of Paris. As he reorganized his units, Foch had a new priority, one forged in adversity: 'Infantry was to be economized, artillery freely used and every foot of ground taken was to be organised for defence.' Morale was important, but to secure enduring success an attack had to be backed up by overwhelming firepower, have a decent numerical superiority and the troops had to be ready to resist the inevitable German counter-attacks.

The crunch came as the Germans' wheeling attack began to move in ready to attack Paris from the east. The battle that resulted, on 6 September 1914, was a huge sprawling affair, but as part of the greater scheme devised by the French commander-in-chief, Joseph Joffre, Foch was ordered to attack the flank of the German advance, feeling for the gaps between the German First and Second Armies. The fighting was desperate, and at one point it appeared that Foch's Ninth Army was about to be overrun. His response in adversity has become a legend: 'My centre is giving way, my right is in retreat. Situation excellent. I attack.' This superficially seems to hark back to his old theories, but this time Foch's apparent folly was underpinned by the knowledge that if his defence was crumbling and retreat was impossible, then in such a decisive battle there was nothing left to do but to attack. He had to do something to derail the German plans, and he had the inner steel and conviction to order his men forward come what may. In the end it was a close-run thing, but Joffre's plans came to fruition and the exhausted Germans were forced into the retreat. France had suffered grievous losses and so too had Foch, whose own son had been killed in the fighting. His reaction was typical of the man: 'I can do nothing more for him. Perhaps I can still do something for France.'

Hero of Ypres

After the Marne, Foch was appointed as assistant chief of staff and given command of the Northern Army Group. Both sides were trying to outflank each other, in what would become known as the 'Race to the Sea'. Here Foch had to learn the difficult business of alliance warfare, for he was responsible for the front also occupied by the British Expeditionary Force (BEF) and the Belgian army. Unlike many of his contemporaries, Foch recognized that diplomacy and an element of compromise were essential for success in such circumstances. He soon obtained the cooperation of Field Marshal Sir John French, commander of the BEF, and funnelled in the French reinforcements needed to help the struggling British hold the line at Ypres. Foch realized that further retreat was impossible, since the vital Channel ports lay just a few more miles behind Ypres. The BEF had to hold and Foch made sure that they did. His interventions were crucial as he prevented a split developing between the French army and the BEF. The putative line around Ypres contracted as the Germans battered away, but it managed to hold. By the end of November the Western Front stretched from Switzerland to the North Sea. The Germans would launch one more offensive on the Ypres Salient on 22 April 1915, this time backed by the release of clouds of chlorine gas. Once again the Allied lines held, and from then on the German army was committed to defending the great strip of France and Belgium they had gained.

A brick wall: 1915–16

By this time Foch had grasped that attacks on German infantry dug in and supported by machine-guns and artillery would not usually succeed. This was even more the case when the original shallow-dug ditches mutated into multi-layered trench systems. Yet if they did not attack then the Germans would remain in situ, and that too was unacceptable. The Allied generals found themselves unable to overcome these problems with the weapons and resources at their disposal in 1915–16. They tried to use artillery power as a bludgeon to smash their way through, but it was impossible to amass the numbers of guns, shells and trained gunners required. In truth, new concepts and new weapons were required before it would be feasible to break through. Faced with these intractable problems Foch fared no better than any of his contemporaries – they all struggled to master the new language of war. The May 1915 attacks made by Foch's armies in Artois were dreadfully expensive in lives and had negligible results. The Allies again tried attacking in the Artois, Loos and Champagne areas in September. Again they got nowhere very slowly. Allied offensive tactics were simply not working in 1915.

In February 1916 the German army launched an offensive on Verdun with the murderous intent of sucking French reserves into the maw of a German mincing-machine of massed artillery. The French could never surrender the historic and deeply symbolic fortress of Verdun, and the darkest attritional battle of the Great War began (see pp. 84–6). For once the Germans had miscalculated, for they too were sucked into the mêlée and they too suffered excessive casualties. The battle would drag on for eleven months.

Foch was not involved in the Battle of Verdun, but he was responsible for French participation in the offensive to be launched alongside the British in the Somme area on

1 July 1916. This was originally designed to achieve breakthrough, but became, at least in part, a campaign to alleviate the German pressure on Verdun. By now Foch's reliance on artillery rather than élan was explicit. Victory now depended on 'a series of successive acts, each one necessitating a great deal of artillery and very little infantry'. Yet the Allies still did not have enough heavy guns, or the artillery techniques to secure tactical surprise. Most significantly, they still did not understand that it was far more efficient to suppress the ability of defenders to fire back during the moments of assault than to try to destroy them. The Somme offensive proved a disaster for the British; but the French started off reasonably well, only to get bogged down in yet more attritional fighting.

By this time the French politicians had had enough of the huge national sacrifices demanded by Joffre: he had more than exhausted the credit he had earned on the Marne in 1914. When Joffre fell, Foch fell with him.

The all-arms battle

While Foch's star was temporarily eclipsed, it was others who slowly, painfully, felt their way to the solution of the problems of successfully attacking on the Western Front. Munitions workers sweated to produce the requisite guns and ammunition; new technologies allowed the gunners to identify the position of their targets exactly and to open fire accurately without the prior registration that surrendered surprise; gas shells could saturate German batteries to suppress any return fire; the new tanks could crush the German wire under their tracks and take out German machine-gun posts; aircraft sped low over the battlefield, attacking targets of opportunity. Above all, the infantry had learnt that discretion must temper natural élan. They now felt their way forward covered by additional firepower generated by light machine-guns, mortars and showers of hand grenades. This was the 'all-arms battle', in which flesh and blood was replaced with machines. Firepower was the key.

Foch returns

In 1917 the collapse of the Russians on the Eastern Front allowed the Germans to concentrate their forces for a last attempt in the spring of 1918 to break the strategic deadlock in the west. Their window of opportunity was short, for the American army was slowly gathering its strength ready to intervene on the Western Front in midsummer. Foch had been brought back as chief of staff of the French army, serving alongside the new French commander-in-chief, General Philippe Pétain. Foch's role was crucial, not only in rebuilding the shattered morale of the French army broken in the Champagne battles of April 1917, but also in putting into place the huge stockpiles of munitions that he knew would be needed for a successful offensive. He was also increasingly involved in Allied grand strategy, and indeed was beginning to be mooted as a possible future supreme commander. As a first step he was placed in charge of the theoretical general reserve set up by the Allied Supreme War Council.

Foch takes command

On 21 March 1918 the Germans hurled themselves forward, striking hard at the British in the Somme area. The badly outnumbered British fell back, and for a while it seemed that the Germans would succeed in driving between the British and French armies. Both Pétain and Douglas Haig, the commander of the BEF, were mutually suspicious that the other was about to revert to selfish nationalistic interests: the French to fall back and cover Paris; the British to cover their retreat to the Channel ports. At an emergency Allied conference convened at Doullens on 26 March, Foch was widely acclaimed as the perfect man to coordinate the desperate defence. He had done it before on the Marne and at Ypres – he could do it again. And he did. His stubborn determination stiffened resolve, and strategic priorities were firmly reasserted. He controlled the Allied reserves with impartial severity, doling them out only as absolutely necessary and generally husbanding them for use when the tide should turn – as he knew it would. With increasing desperation the Germans launched more massive offensives, each time gaining ground, but failing to break the integrity of the Allied line. Soon they were left with a collection of tumescent bulges pointing the way to nowhere. The German army was exhausted and it was plain that these ungainly salients were all too vulnerable to counter-attack.

Advance to victory

When he was appointed as Supreme Allied Commander-in-Chief, Foch immediately pulled the Allies together. He was almost alone in seeing the armies fighting on the Western Front as one force, to be directed in a single plan. By the end-phase of the Great War he was operating at the level of grand strategy, moving his armies across the western European battlefield in giant sweeps, with tactical geographical considerations almost irrelevant to his greater design – the total destruction of the military capacity of imperial Germany.

After the failure of the last German offensive in July 1918, Foch knew that the Germans were exhausted – and he knew that this was the moment for his old theories of the offensive to be dusted off. He responded by launching a sustained onslaught that matches anything else in military history. His most intuitive skill as a master of alliance warfare was to apply the right army to the task in hand: the French army was still teetering on the brink of exhaustion and the Americans were numerous but far too raw and inexperienced. So it was that Foch clearly perceived that Haig's BEF would have to be his main strike force. After painfully learning the grim trade of war, the British had fully grasped the concept of the all-arms battle, and it is to Foch's credit that he did not shrink from giving them the lead role in the 'Hundred Days' series of offensives between 8 August and 11 November. He ordered the British forward, time and time again, until their battalions were worn down to the bare bones. At the same time he skilfully wove in renewed assaults by the French army and launched forward the new American armies.

Foch drove his own generals as hard as he drove the Germans. He never stopped badgering them, always asking for more and demanding they take risks that would once have been considered suicidal. By switching focus from one end of the front to the other he never gave the German High Command a moment's peace to sit back and properly reorganize their defences. German morale wavered and collapsed, and the result was an utterly comprehensive Allied victory. At every level the German war machine was beaten: their leaders were utterly demoralized, the much-vaunted storm troopers ground to dust, the elemental power of the German artillery quelled and the final victory secured over the solid phalanx of infantry divisions that had borne the brunt of the fighting and had at times nearly – but not quite – won the war for Germany. This, then, marked the final triumph for which Foch had striven all his adult life.

Armistice and after

When it came to the Armistice negotiations, Foch remained true to his beliefs. To ensure that the Allies could impose their collective will on the Germans, he sought conditions of such severity that the German armed services on the ground, at sea and in the air were stripped naked before their enemies. There was no way that the Germans could feasibly resume the war.

Foch was determined to hammer the Germans down into their coffin, and became estranged from the French politicians who sought to draw a line under the Great War. He was, however, acclaimed on both sides of the Channel as one of the key Allied architects of victory, and was made a British field marshal in recognition of his services. He died on 20 March 1929.

PHILIPPE PÉTAIN

1856–1951

CHARLES WILLIAMS

MARSHAL PHILIPPE PÉTAIN was one of the greatest – and least appreciated – masters of warfare in the First World War. Alone among the generals on all sides, he quickly understood the battle-field effects of heavy artillery, barbed wire and machine-gun. As a consequence, he developed, as early as the autumn of 1914 and, more potently, at Verdun in 1916, tactics designed, above all, to mitigate the resulting casualty rate. It was this concern for the men under his command that allowed him to settle the mutinies of 1917 and, in the end, to bring the French army to the point where it was the finest fighting force of its time.

Pétain's reputation, even as a general, has been blackened by his acceptance of defeat in 1940, by his record as the leader of the collaborationist Vichy government, and by his subsequent conviction for treason (in a travesty of a trial) in July 1945. Nevertheless, whatever the negatives, he is still known as the 'Victor of Verdun'. True, his supporters cannot claim that he possessed the speed of thought of a Napoleon Bonaparte or the fierce determination of a Wellington, but in the understanding of the human element of warfare, and also in tactical intelligence, he ranks not far below them.

The son of my sorrow

Henri Philippe Bénoni Omer Pétain was born on 24 April 1856 into a peasant family in the northern French department of Pas-de-Calais. When he was 18 months old, his mother died (it was she who had called him 'Bénoni', meaning the 'son of my sorrow'), and on his father's remarriage he was farmed out to his grandparents. More attractive, however, was the home of his uncle, the priestly incumbent of the neighbouring village of Bomy. There Pétain was able to listen to extravagant military tales from his great-uncle, by then also in holy orders, who had fought in Napoleon's army in Italy. When it came for him to choose a future career, between peasant farming, the Church or the army, there was simply no contest. It would be the 'glorious' French army (in fact, only just recovering from its disastrous defeat in the Franco-Prussian War of 1870–71).

The long apprenticeship

The decision taken, his clerical uncle was clear on one matter: the boy should not just go into the ranks but should join as an officer. Accordingly, the young Pétain went through the educational steps necessary to take him to the École Spéciale Militaire (Saint-Cyr) at Versailles. But he did not do well there, ranking, on merit, 229 out of 386. The cavalry, his original ambition, could no longer be a target. That left the infantry, where the choice was between the colonial army and the domestic army. Although he knew that promotion was quicker in the colonial army (officers were killed at a greater rate), he opted for the domestic army, and was posted to a battalion of the Chasseurs Alpins stationed in the Mediterranean fishing port, as it then was, of Villefranche-sur-Mer.

There was not much to do there, apart from seducing the wives of local dignitaries (at which the good-looking Pétain became an expert). Nor, for that matter, was there much else to do as he slowly crawled up the ladder of peacetime promotion; Besançon in 1883, where he made an unconvincing effort to get married, only to be refused by his prospective parents-in-law on the grounds of his peasant background; the École Supérieure de Guerre (better known as the École de Guerre) in Paris in 1888; a position as staff officer in Marseille in 1890; the command of an infantry battalion at Vincennes in 1893; and then back to Paris in 1895 on the staff of the military governor, carefully avoiding there any involvement in the topic of the day – the Dreyfus affair.

Pétain's lectures

One of the advantages of being an officer on the staff of the military governor of Paris was the possibility of an introduction to the higher-ranking officers in the capital's military establishment. Thanks to just such an introduction, Pétain was invited in 1901 to lecture at the École de Guerre. Yet it was not until he returned as a full professor in 1908 that his lectures provoked serious controversy, particularly in 1910, when he found himself at odds with Lieutenant Colonel Louis de Grandmaison, who was loud in advocating the doctrine of *offensif sans arrière pensée* ('offensive without second thoughts'). Petain was more cautious, arguing for intensive preparation of the ground by artillery and machine-gun firepower – then, and only then, to be followed by infantry attack.

Where Pétain ran into trouble was in the presentation of his views. He seemed to be advocating a pause before serious engagement. Grandmaison countered that any pause would allow the enemy to regroup. But although much was made at the time of the supposed clash of opinion between Pétain and Grandmaison, there was, in fact, no great clash. Grandmaison was dealing with the deployment of whole armies, while Pétain was dealing with infantry tactics in localized battles. Nevertheless, in the popular mind – there was much debate on the matter in the Parisian press – Pétain was tarred with the brush of 'defensive' while Grandmaison was lauded as 'offensive'. As a result, Pétain was told – informally, of course – that he had reached the end of his military career and could expect no further promotion.

1914: retirement postponed

Pétain duly planned his retirement. He would, after all, be 60 in 1916, the retiring age for brigadiers. He found a plot of land in the Pas-de-Calais, and even bought a pair of secateurs with which, as an old retired bachelor, he would cultivate his garden. In April 1914 he left the command of the 33rd Infantry Regiment – where he had welcomed the arrival of one of his greatest admirers, the tall young sous-lieutenant Charles de Gaulle – to take over the 4th Infantry Brigade, headquartered at Arras. Its parade at Arras on Bastille Day, 14 July 1914, was meant to be his last show-piece. He rode on a white horse on to the parade ground at a brisk trot, inspected his brigade, turned his horse round and went off at the gallop.

One month later, of course, retirement was indefinitely postponed. Pétain then led his brigade into battle. During the Battle of the Frontiers in late August and early

September 1914 he and his troops went on a series of forced marches into Belgium and back again, ending up, after a series of bruising engagements, on the River Marne, fighting to halt the German advance. There Pétain caught the attention of the French commander-in-chief, General Joseph Joffre. In fact, Pétain was one of the survivors, and indeed one of the beneficiaries, of Joffre's subsequent purge of obviously incompetent generals, and in October found himself in command of the 33rd Army Corps at the northern end of the French line of battle, in front of the city of Arras.

The battles of Artois, Champagne and Verdun

Pétain hardly had time to settle in before he was called on to repel a German attack aimed at Arras itself. By the time he took up his command, German advance units were within 2 miles of the outskirts of the city. After some desperate fighting the enemy attack was halted and an (unsuccessful) counter-attack launched. The pattern repeated itself for several weeks, but, as time and casualties wore on, Pétain realized that the tactics were mistaken. A strong defensive first line and a relatively weak second line, the official doctrine of the day, was the wrong way round. Much better was a weak first line and a strong second line out of the range of heavy artillery. Moreover, even in the first line it was no good to have men just standing waiting to be shot. Pétain put the emphasis on camouflage and concealment, with centres of resistance, for instance in farm buildings, rather than on a line of poorly protected men on open ground. Trenches in the second line were dug deeper and machine-guns more carefully hidden. It was only when the enemy attack had been absorbed and weakened by the second line that a counter-attack could take place.

Pétain's tactic worked. Successful defence ensured that the Battle of Artois ground to a stalemate. All this, of course, was duly reported to Joffre. Joffre, however, insisted on an all-out offensive in the spring of 1915. Dutifully obeying orders, but with reluctance, Pétain launched the offensive on 9 May. At first, all went well. By mid afternoon his infantry had advanced to capture the commanding feature of Vimy Ridge. There was much congratulation (Pétain was made a commander of the Légion d'Honneur in Joffre's orders of the day), but, in fact, the Germans had adopted Pétain's own tactics and had moved their reserve divisions into a strong second defensive line. The French attack stalled and was then beaten back. By late June, when Joffre finally called off the offensive, no ground had been made and the French had lost 100,000 men.

By then, Pétain had been promoted again, to command the French Second Army, holding an extended line north of Châlons-sur-Marne in the Champagne. There again, after some bizarre tricks to try to keep his arrival a surprise for the Germans (his reputation had gone before him), the order came from Joffre to launch an offensive in the autumn to coincide with a British attack in the north on the village of Loos. Again, with reluctance, this was done.

In the event, all the attacks failed. The Germans had fully learnt the lesson of a strong defensive second line (particularly if invisible to the attackers on a reverse slope) and, true to Pétain's own doctrine, had absorbed the French attack at the second line and regained the ground lost in the initial assault. By the time the Champagne offensive was called off by Joffre at the end of October 1915 the official French losses were nearly 200,000 men.

Yet again, Pétain reported that such general attacks were a waste of men and ammunition. The answer, he argued, was a tactic of limited strikes at the weak enemy front line and a quick retreat – in the hope of provoking him into a hasty counter-attack. In response, an irritated Joffre posted Pétain to direct the training of four army corps of reservists ready for another major spring offensive in 1916. Thus relieved of his command, Pétain took up his new post in the relative tranquillity of Noailles, just outside Paris. It was frustrating, but there was good food and wine, and many assignations with his mistress in the Hôtel Terminus at the Gare du Nord in Paris. Such were the pleasures of this life away from the front that it came as little more than an irritation to hear that on the morning of 21 February 1916 the German heavy guns had at first light begun their barrage at Verdun.

Almost immediately Pétain was summoned by Joffre from one of his assignations at the Hôtel Terminus to take command of the defence of the Verdun salient. By a series of imaginative tactical moves he managed to save a situation that appeared to be lost and turn it into a platform for attack.

Verdun

By the end of 1914, with the stabilization of the Western Front, the Verdun salient had become the lynchpin of a defensive line stretching from the Channel to the Swiss frontier. The city itself, standing astride the River Meuse, was at the neck of the salient, protected by a ring of forts (although denuded of their heavy guns). Late in 1915 the German High Command decided to pre-empt an expected Franco-British assault in the northwest on the Somme by attacking Verdun. The German tactical build-up during January and February 1916 was massive. Ten new railway lines were constructed. Some 1,400 guns, 2.5 million shells, 168 aircraft and 400,000 men were delivered to the front. To cap it all, the Germans had a terrifying new weapon – the flamethrower.

On 21 February the German heavy artillery opened up with a barrage aimed at the French positions on the eastern, or right, bank of the Meuse. This was followed, just as the winter dusk was falling, by groups of German storm troopers, not in the customary waves but in zigzagging, crouching runs. The flamethrowers wrought terrible and fearsome havoc. In spite of localized and fierce French resistance, in the first week the fort at Douaumont fell, and the Germans advanced 5 miles into open ground – only 2.5 miles from the last line of defence in front of the city itself.

Pétain was brought in to assume command of the French defence on 26 February. He immediately ordered a tactical change: the defence would now consist of a forward line designed to blunt an attack but not to be held at all costs, a principal 'line of resistance' to be held at all costs, and a third line from which counter-attacks would be launched. He also organized the 'millwheel', a system of constant replacement of units in the front line with fresh units. (Fresh troops and ammunition were carried along a supply route between Bar-le-Duc and Verdun that came to be known as *La Voie Sacrée* – 'the sacred way' – and which has found its place in the canon of French military heroism.) Halted by these tactics on the right bank, Crown Prince Wilhelm, the German field commander, attacked at the beginning of April on the left bank of the Meuse, but after more prolonged fighting this too was held.

At the end of April Pétain was promoted to command of the army group in the Verdun sector, while General

Robert Nivelle took the immediate field command – although Pétain made daily visits to his headquarters. A further German offensive in June threatened again to break through but was stopped in front of the fort at Souville. It was to be the final German effort. In October the French, led by Nivelle and the ferocious General Charles Mangin, counter-attacked. By December they had recovered virtually all the ground lost. The whole exercise had turned out to be fruitless. By way of footnote, French casualties during the dreadful ten months of the battle were estimated at 550,000, German casualties at 434,000. Much to Pétain's disgust, Nivelle then jumped above him to become commander-in-chief of the French army.

The 1917 mutinies

Nivelle's 1917 spring offensive was a disaster, as Pétain had predicted. Not only did it fail in all its objectives but it destroyed the morale of the French army. Mutinies broke out along the line in April and gathered pace in May. In a panic the French government sent for Pétain to replace Nivelle as commander-in-chief and to settle the mutinies. Pétain agreed, but only if he could do it in his own way and in his own time.

Unlike other generals before him Pétain visited every unit in the front line to hear the men's grievances at first hand. The main problem, as he found out, was that the soldiers had lost contact with their families, since they had not had any leave. Pétain immediately ordered that seven days leave every four months should be mandatory, at the same time arranging for extra trains and seeing to it that reception arrangements at Paris terminals were improved, with the French Red Cross providing canteens for arriving soldiers. He also demanded an improvement in the quality of men's rations in the front line. By July the mutineers had decided that they had made their point. (Not that Pétain was soft: some 49 men were shot and nearly 1,400 sentenced to deportation or forced labour.)

The offensives of 1918

Pétain then set about rebuilding the French army. A series of successful limited engagements in the autumn helped morale. So did the arrival of General John Pershing and the advance units of the American army. It was by then obvious that it was only a matter of time before the Americans would appear in force. Everybody knew that – not least General Erich von Ludendorff. In March 1918, in an attempt to finish the war before the Americans arrived, Ludendorff launched his last offensive. It was almost a complete success. The Germans broke the British Fifth Army and threatened a breakthrough to the Channel. The British commander-in-chief, Sir Douglas Haig, appealed to Pétain for help. Pétain immediately sent fifteen French divisions to seal off the breach in the British line in front of Amiens and promised twenty-five more when they could be mustered. But in doing so he weakened his own defensive positions to the south. Ludendorff then diverted his attack there in late May, to open up the way to Paris. Pétain retreated – and went on retreating until he could be sure that the German supply lines were over-stretched. On 18 July he ordered the counter-attack. The German offensive had been halted – and Paris was safe.

That done, in October 1918 Pétain hatched a further plan. Starting from the northeast of Nancy in Lorraine, he and Pershing would strike eastwards into Germany beyond the River Saar, turn south and envelop the remaining German army on the Western Front, forcing a humiliating surrender. But the plan came too late. An armistice had been decided. On hearing the news, Pétain, as he said himself, was in tears. The opportunity to destroy the German army once and for all had been lost.

From Verdun to Vichy

Pétain lived for the rest of his long life in the shadow of his successes of 1914–18. Awarded the marshal's baton in December 1918, he outlived the other six marshals of the war. His last military command was in Morocco in 1925, where he met the Spanish Falangist leader Primo de Rivera and, like many generals before and since, became more and more authoritarian in his views. The result was Vichy, conviction for treason and a death sentence, commuted to life imprisonment. He died in prison in 1951.

But that was not altogether the end. Successive French presidents, from de Gaulle in 1968 until Jacques Chirac in 1998, arranged for a single red rose to be placed on his grave each year on 11 November – the anniversary of the 1918 Armistice. In spite of the shame of Vichy, the 'Victor of Verdun' had not been entirely forgotten.

EDMUND ALLENBY

18261–1936

JEREMY BLACK

THE FIRST WORLD WAR is not noted for successful offensive operations, particularly on the part of the Entente Powers, but Edmund Allenby achieved two such. The first, in late 1917, led to the capture of Jerusalem from the Ottoman empire, while the Battle of Megiddo in late 1918 overthrew the Turkish position in Palestine and Syria and brought the war in the Middle East to an end. Subsequent difficulties in campaigning in the region have highlighted Allenby's achievements. They brought him fame and fortune and established Britain as a land power in the Middle East. Allenby's success contrasted with the inability of General Archibald Murray, his predecessor as commander of the Egyptian Expeditionary Force, to defeat the Turks.

The eldest son of a wealthy British landowner, Edmund Henry Hynman Allenby was educated at Haileybury and sought a career with the Indian Civil Service. Failure twice in the entrance exams led him to turn, instead, to the army, and he passed out of the military academy at Sandhurst in 1881. A keen countryman, Allenby then took a decision that was to be very important in his military career: he joined the cavalry.

His regiment, the 6th (Inniskilling) Dragoons, was based in South Africa, and he joined it there in 1882. Aside from a two-year break, from 1886, at the cavalry depot in Canterbury, Allenby served in South Africa until the regiment returned to Britain in 1890. This service did not entail fighting, as he arrived too late for the Zulu War of 1879, but Allenby, eventually as adjutant, became an experienced officer, well able to execute tasks and take responsibility. However, he also became a somewhat harsh figure, focused on discipline, and this was to remain a characteristic of his command style.

In 1896, the year of his marriage, Allenby entered the Staff College at Camberley, where his class included Douglas Haig, the British commander on the Western Front from 1915 until 1918. Competent, not brilliant, at Staff College, Allenby benefited from his stint by becoming first a major (1897) and then adjutant to the 3rd Cavalry Brigade (1898).

The Boer War and after

Not a triumph for Britain, the Boer War with the Afrikaners of the Transvaal and the Orange Free State in southern Africa (1899–1902) nevertheless made some reputations as well as destroying others. Allenby benefited from his ability to act as an effective cavalry commander in the campaigns to drive the Boer opponents from the veldt. He spent many hours in the saddle and his physical fitness was crucial to his ability to fulfil his command responsibilities. After being given temporary command of his regiment in 1900, he became a lieutenant colonel.

Following the war, Allenby continued his steady rise, appointed brigadier general (1905), major-general (1909), and then inspector general of cavalry (1910). The last brought out his capacity to act as a martinet. He was certainly not associated with any particular insights as a military thinker.

The Western Front

The outbreak of war in 1914 brought command, first of the Cavalry Division and then of the Cavalry Corps. Like his colleagues, Allenby struggled to respond to the massive German attack. His defensive role in the First Battle of Ypres (1914) was effective in helping anchor the British front, and, in 1915, he was promoted to command, first of 5th Corps and then of the Third Army.

In the latter capacity, Allenby directed the unsuccessful Gommecourt diversionary attack at the northern end of the Somme offensive in July 1916. The following April, his reputation was seriously compromised in the attack at Arras. This became an attritional offensive, in which Allenby failed to show the necessary flexibility in command, leading to complaints from three divisional commanders. In the face of such complaints, he could no longer remain in command in France. This was the background for his dispatch to the Middle East.

The problem at Arras, as with other battles in 1917, was the inability of the British to break into the full depth of the German defences, consolidate and press home any advantage that arose, so that break-in could be converted to break-through. German counter-attacks had to be neutralized, but artillery and trench mortar tactics had yet to become sufficiently developed at interdiction, although Arras, in its turn, helped provide the lessons from which better tactics developed. The inability to exploit success was as much a problem of communication and control as it was one of planning, stamina and reinforcement.

Allenby came in for criticism twice over Arras: first, for wanting too short a preliminary bombardment, and later for not ending his assault when it had clearly ground to a halt. Whether the latter was due to his lack of awareness of the true situation because of poor communications, his stubbornness or a genuine belief that he could achieve his objective if he kept pushing, is not clear. He had an aversion to heeding what he thought was bad advice. His performance at Arras seems to have been used as a reason to get rid of him, but he was no worse than Hubert Gough, who commanded the Fifth Army. Moreover, the success of the Canadians at Vimy, which coincided with the start of the Arras offensive, had overshadowed Allenby's initial achievement there.

In the Middle East, 1917–18

Allenby regarded his dispatch to replace the dismissed commander of the Egyptian Expeditionary Force (over which he assumed command on 28 June) as a mark of disfavour; but he transformed the situation there. His diligent oversight of his new command exuded vigour and raised morale. Moving his headquarters closer to the front was important both in practical terms and as a morale-boasting move.

Allenby, however, faced the serious problem of coping with very different expectations from his superiors. David Lloyd George, the hyperactive prime minister, wanted bold offensive thrusts and the capture of Jerusalem to help morale; while William Robertson, the Chief of the Imperial General Staff, favoured a more cautious approach. Recent failures at the hands of the Turks in Mesopotamia

(Iraq) and at Gallipoli served as stark warnings of the dangers of boldness.

Allenby's Boer War experience of working with Australian troops served him in particularly good stead, as his new command was very much that over an army of the British Empire. Having revived the army, which he strengthened with troops, heavy artillery and aircraft, Allenby decided to implement the plan for a new offensive that had been already prepared by his new staff, particularly Lieutenant General Sir Philip Chetwode.

The resulting Third Battle of Gaza led to a shift in the axis of attack, away from Gaza and the coast (the axis in the costly first two battles in the spring of 1917), towards the eastern end of the Turkish positions near Beersheba 30 miles inland. This was captured by the Australians on 31 October after a long advance across terrain with very little water, an advance that required intelligence, reconnaissance and engineering work. This was a surprise attack launched without a prior bombardment, and the Turks were unable to destroy the town's water wells as expected. Cavalry helped provide the British with the necessary mobility, but flexibility of mind was also required to vary the direction of attack. Near Gaza, moreover, where the attack began on 27 October, Allenby's forces used tactical lessons from the conflict on the Western Front to help break through the Turkish positions.

Victory was exploited with the capture of undefended Jerusalem on 9 December. Earlier, there had been tough fighting in the hills of Judea, and the British suffered 18,000 casualties in the campaign. The outnumbered Turks suffered heavier losses. The fall of Jerusalem helped to catapult Allenby into prominence. On 11 December, he publicly entered Jerusalem in a carefully scripted display. Rather than riding in, as Kaiser Wilhelm II had arrogantly done in 1898, Allenby dismounted and entered on foot. In a season without good news for the British public, Allenby's success was very welcome to the government, which therefore devoted considerable attention to it. The Turks, however, who had not been crushed, regrouped on a new defensive line to the north of Jerusalem. Their ability to do so has been ascribed by some critics to overly cautious generalship on the part of Allenby, but this was in keeping with his preference for methodical warfare, a preference towards which strong logistical pressures powerfully contributed. There was not enough water for the cavalry, while the strength of the opposing rear-guard was also a factor. Moreover, the prospects for campaigning were affected by the winter rains. The Cabinet wanted Allenby to press on for Damascus, but he urged the need to wait until the necessary infrastructure, including a double-tracked coastal railway, was ready.

Allenby, nevertheless, pressed forward to attack the

Turks, capturing Jericho as part of an advance into the Jordan Valley, which put pressure on the Turkish flank. However, he was held back by the extent to which his army was used to provide reinforcements for the British forces in France, which were under heavy pressure from the German offensive in the spring of 1918. Fifty-four battalions were transferred. Their poorly trained Indian replacements took a while to adapt, and Allenby was concerned about Turkish attempts to woo Muslim sentiment in their ranks.

In the spring of 1918, Allenby launched two unsuccessful operations to the east of the Jordan. They were designed to cut the Hejaz railway from Syria to Medina and Mecca, and thus to support the Arab revolt, both by hitting Turkish communications and by creating a supply route for the Arabs from Palestine, so that they could advance into Syria. The first raid, however, was affected by the winter rains, which made it difficult to move the artillery. As a result, Amman was not captured. A threatened Turkish counter-attack led to the abandonment of the second raid.

The summer provided a welcome opportunity to develop the army, through the introduction of improved weaponry and the creation of a better logistical structure. A Turkish attack at Abu Tulul was defeated in July, and in September Allenby won a decisive victory at Megiddo.

On 20 September 1918, Allenby launched the final offensive (with 69,000 troops and 550 pieces of artillery) against the Turkish forces (34,000 and 400 pieces) in Palestine. This battle was called Megiddo as Allenby's troops advanced by this ancient mound, the alleged site of the final battle (Armageddon) mentioned in the biblical Book of Revelation. Tanks, aircraft and cavalry were all vital, although, as on the Western Front in 1918, an effective artillery–infantry coordination was important in breaking through the opposing lines. Furthermore, Allenby's skilful strategy kept the Turks guessing about the direction of attack. He began by raiding the eastern end of the Turkish line, creating the impression that he would repeat his 1917 plan; but, in the event, did the opposite.

Allenby attacked with his left near the coast. The breakthrough with infantry and artillery provided an opportunity for the British cavalry to exploit as the cavalry had not been exhausted in the initial assault. They swung to the east, cutting the line of Turkish retreat. At the same time, other cavalry units advanced along the coast, capturing both Haifa and Acre on 23 September.

This dramatic success, in which 75,000 prisoners were taken for 5,666 casualties, was a triumph for mobility. Megiddo is widely seen as the last great cavalry battle. The Australian Light Horse used cavalry charges as well as acting as mounted infantry. Other weaponry, however, was also crucial. For example, armoured cars provided significant mobility. British forces in this context should, throughout, be understood to include imperial troops, in large numbers, particularly from Australia, India and New Zealand. Allenby's use of subterfuge to mislead the Turks and his planning at Megiddo show that he was open to unconventional ideas and that he did not underestimate the Turkish forces, even though they were under-strength and badly equipped. It is interesting that this kind of misdirection was again used successfully in the Second World War, particularly by Montgomery at El Alamein in 1942, and prior to D-Day in 1944, both meticulously planned operations.

Allenby's use of air power in Palestine was also highly effective. Only air supremacy (as opposed to air superiority) could prevent the German air contingent supporting the Turks from flying reconnaissance sorties. Allenby, though, had an accurate picture of the Turkish position. Air supremacy also prevented interception of the RAF's bombers, which were used to bomb telephone exchanges and rail junctions to disrupt communications and to destroy the Turkish forces when they retreated. (Air power was used in Normandy in the same way in 1944.) There was even naval support from two destroyers at Megiddo.

Allenby's mobile forces included 12,000 cavalry, Prince Faisal's Arabs, and armoured cars, all of which he used very effectively to break through, encircle and trap the Turkish forces. Their destruction with the aid of air power presaged the same tactic over northwest Europe in 1944 by the Second Tactical Air Force. It is noteworthy that the artillery bombardment at Megiddo only lasted 15 minutes, showing that Allenby knew the value of neutralization and shock. He was able to master the factor of time. A rapid advance on two axes then further exploited the situation. Along the coast, troops advanced to capture Tyre, Sidon and, on 2 October, Beirut. The previous day, Australian cavalry that had advanced from the Sea of Galilee joined Arab forces (advised by T. E. Lawrence) in taking Damascus. Syria was rapidly conquered, with Aleppo falling on 25 October. Five days later, an armistice signed at Mudros ended the conflict.

Allenby was helped by the range of Turkish commitments, as well as by the extent to which Turkish planners were more interested in exploiting the Russian Revolution of the previous year to make gains in the Caucasus, where they thus concentrated their forces. As a result, the Turks in Palestine were short of men and equipment and could not match the British, who outnumbered them and had air superiority. The demoralized Turks, their coherence fractured by the rapid British advance, readily surrendered, although some units mounted rearguard actions.

Allenby's victory helped ensure that Britain became

the key power in the Middle East, which was seen as an important imperial goal. In the short term, this gave the British a strong negotiating hand in the post-war Paris peace talks, not least since the French wished to become the colonial power in Syria and Lebanon.

Assessment

As a general, Allenby was at his best when he was allowed to plan and fight without interference from above. At Arras, he had to satisfy Haig, whereas in Palestine he was more his own boss, albeit still answerable to meddlesome political and military oversight from London.

Allenby had several qualities that suggest that his generalship was sound. He was concerned for the men under his command, and well aware of the importance of good morale. He was meticulous. When he saw opportunities, he tried to take them. His use of cavalry was always good. He always wanted to fight on terms that he dictated to the enemy. His use of air power was very good. However, against these qualities, he had an explosive temper which suggests that he expected full compliance with his instructions and would not listen to subordinates. The implication is that he was less willing to let others have their way or use their initiative. He also had a reputation for being a martinet, which cannot have endeared him to his men.

Palestine allowed broad sweeping movements by

mobile forces. Nevertheless, Allenby played to his strengths and minimized his weaknesses by not attacking in the spring and summer of 1918 when his experienced troops had been sent to France because of the German offensives.

After the war

The end of the war brought prestige, with Allenby becoming 1st Viscount Allenby of Megiddo and Felixstowe (the title linking the decisive battle and the Allenby family home), as well as a field marshal. He was also voted £50,000 by Parliament.

Allenby himself continued to be linked to the Middle East. From 1919 until 1925, he served as special high commissioner for Egypt, a key posting for imperial security. As such, he had to use troops to suppress the Egyptian rising of 1919. He had greater success than his counterparts in Iraq, let alone those responsible for the unsuccessful British intervention in the Russian Civil War, and did not use the heavy repression shown by the French in Syria in the mid 1920s. Alongside his role in counter-insurgency, Allenby sought to consolidate the British position by appropriate policies. He recommended the end of the protectorate that had been declared in 1914, and in 1922 Egypt was given a greater degree of independence, although with Britain retaining control over security and foreign policy as well as a large garrison in the Suez Canal. After retiring in 1925, he focused on bird-watching, fishing, travelling, and his presidency of the British National Cadet Force. He died suddenly in London, on 14 May 1936.

JOHN PERSHING

1860–1948

CARLO D'ESTE

GENERAL JOHN JOSEPH 'BLACK JACK' PERSHING was the most famous American soldier of his era. During a lifetime of soldiering, he rose to become the only officer ever to be accorded the title of General of the Armies. As the foremost American officer of his time, during the First World War Pershing commanded the American Expeditionary Force (AEF). Pershing's earlier colourful and star-studded career had taken him from the western plains during the Indian Wars, to the Spanish–American War as a Rough Rider under Teddy Roosevelt, where he won a Silver Star at San Juan Hill, and from fighting insurgents in the Philippines to a hunt for Pancho Villa in Mexico in 1916.

Pershing's family was of Alsatian origin and once spelled their name as Pfoershing. He was born on 13 September

1860 in Laclede, Missouri, the son of a railway section foreman for the Hannibal and St Joseph Railway and later the proprietor of a general store. Johnny, as he was known as a child, grew up in the shadow of the Civil War. His first memories were as a young child witnessing his home town being raided and his father and other local merchants robbed by a lawless band of Confederate partisans.

Although his father later became a prosperous businessman, the family suffered in the great depression of 1873. While working on a family farm, Pershing also managed to obtain enough education to secure a teaching position at a local school. Like a later young man in rural Kansas named Dwight Eisenhower, Pershing was determined to gain the benefit of a college education. To help make family ends meet, young Pershing attained a teaching position in 1878. The first example of his legendary reputation for no nonsense and stern discipline occurred when he faced down a rowdy group of students, one of whom had openly challenged his authority. Although his first teaching experience was short lived, it imbued the young man with a renewed desire for more education.

Pershing had no ambition for a military career but, when opportunity knocked, he took a competitive examination for West Point and finished first out of sixteen candidates. After several months of tutoring, Pershing entered the United States Military Academy in June 1882.

He was not an especially notable student, had considerable trouble with French and ranked only thirtieth of seventy-seven in the class of 1886. Surprisingly for an officer who would later achieve a reputation as the army's sternest martinet, Pershing accumulated over 200 demerits, but nevertheless achieved the greatest honour bestowed on a cadet when he was appointed first captain of the corps of cadets. In this capacity he commanded the corps when the funeral train of General Ulysses S. Grant, a man Pershing deeply admired, slowly passed by West Point.

As first captain, Pershing was a disciplinarian – an early indication of what he would be as a future army officer and what his official biographer has described as his 'undoubting certainty of duty combined with a glacial self-possession'. A cadet a year ahead of Pershing who later served as an army commander under him in the First World War once remarked that, while Pershing had earned their respect, he never attained their affection – a trait that characterized his entire life.

Early army career

Pershing was commissioned a cavalry officer and assigned to the 6th Cavalry, based at Fort Bayard in what was then the New Mexico Territory. His first years as a cavalryman were marked by participation in various Indian campaigns in the West, which earned him citations for bravery. One was the notorious incident at Wounded Knee, South Dakota, on 29 December 1890, in which the 7th Cavalry massacred some 300 men, women and children of the Lakota Sioux tribe.

Pershing's reflections on Wounded Knee – in which his regiment played only a peripheral role – revealed an unsympathetic attitude towards Native Americans in general and a belief that the US army had through its actions averted more costly and protracted warfare with them.

During this period, Pershing earned recognition as one of the army's best marksmen with both a rifle and pistol. While assigned as the professor of military science and tactics at the University of Nebraska-Lincoln, 1890–95, he organized a military drill company that eventually was renamed the Pershing Rifles in his honour. To this day, the prestigious Pershing Rifles drill teams exist at a number of American universities and among its distinguished alumni are James Earl Jones and generals Colin Powell and Curtis LeMay. Though he never practised, while at the university Pershing also studied law and became a graduate of the class of 1893.

He remained on the lowest rung of the officer corps' ladder as a second lieutenant until 1895. In 1897, Pershing returned to West Point as a tactical officer. His unbending, iron-fisted insistence on adherence to the letter of the regulations bordered on arrogance and left him with the dubious distinction of being the Academy's most unpopular tactical officer. Even worse, the cadets so disliked Pershing that he was once accorded the silent treatment, whereby the corps ostracized him by refusing to acknowledge his presence in the mess and refusing to eat as long as he remained. Pershing had just returned from an assignment in Montana in which he served as a white officer in the African-American 10th Cavalry, one of the now famous Buffalo Soldier regiments. The West Point cadets tagged him with the blatantly racist label of 'Nigger Jack', a denunciation that over time softened into 'Black Jack', the nickname by which Pershing was known for the rest of his life.

The Spanish–American and Philippine–American wars

The advent of war with Spain over Cuba in 1898 saw Pershing back with the 10th Cavalry as its quartermaster and with a brevet commission as a major in the all-volunteer force organized to fight the Spanish. On 1 July, moments before organizing and leading his black troops in an assault on both San Juan and Kettle hills, Pershing barely escaped death after he stopped to salute the commander of the US cavalry, the dashing Lieutenant General 'Fighting Joe' Wheeler. Wheeler was on his horse in the middle of a creek when a shell exploded so close it soaked both officers.

When the Spanish were routed on San Juan Hill by

Lieutenant Colonel Theodore ('Teddy') Roosevelt's Rough Riders and Pershing's cavalrymen, he emerged from the war not only having gained the favourable attention of Roosevelt but also with a well-earned reputation for bravery. In 1919, a year after it was created, he was awarded the Silver Citation Star, the precursor to the Silver Star, the nation's third highest award for gallantry. In 1932, the army retroactively awarded Pershing the Silver Star for his actions in 1898 in Cuba.

Pershing contracted malaria in Cuba and spent a brief period of service in the War Department in Washington, DC until reassigned to the Philippines in August 1899. There, he became deeply involved in suppressing the Moro uprisings. To better understand the problems of pacifying the indigenous Muslim insurrectionists, Pershing diligently studied their culture and language, including the Koran, and formed close ties with some of their tribal leaders. His superior work in expeditions to eradicate a Moro stronghold earned him both a citation for bravery and notice in Washington, where Secretary of War Elihu Root observed to one of his West Point classmates that: 'If your friend Pershing doesn't look out, he will find himself in the brigadier general class very soon.'

In 1901, Pershing returned to the regular army after his Spanish–American War brevet commission was terminated and was promoted to captain. Reassigned to the 15th

Cavalry, he served as an intelligence officer and participated in further action against the Moros at Lake Lanao, Mindanao, at one point temporarily filling the post of camp commander. With a combination of carrot and stick, Pershing opted for persuasion whenever possible, and when that failed, ferociously attacked and overwhelmed the Moros.

Pershing was posted to Washington in June 1903, his reputation so burnished by his accomplishments in the Philippines that Roosevelt, now president, singled him out by name in a speech to Congress. Roosevelt wanted Pershing promoted to the rank of brigadier general, but his petition to the army General Staff was summarily rejected and, to the president's dismay, he remained a captain.

It was while in Washington that Pershing met and married Helen Frances Warren, the daughter of a wealthy and influential Republican senator from Wyoming. They were married in 1905 in a ceremony attended by Roosevelt. Days later, Pershing and his bride departed for Japan, where he took up the post of American military attaché. In this capacity he became an observer in Manchuria of the Russo-Japanese War of 1905.

In 1906, Roosevelt finally achieved a measure of payback over the army by having Pershing promoted from captain to brigadier general over the heads of 800 more senior officers. Pershing was never popular with his fellow officers, and his promotion resulted in cries of favouritism and the start of a smear campaign that endured for some years. Nevertheless, Pershing became the first example of Roosevelt's insistence that promotions be made on the basis of merit.

In 1908, after once again serving as a military observer, this time in the increasingly unstable Balkans, Pershing returned to the Philippines as the military governor of the Moro province in Mindanao, and until 1913 used a combination of diplomacy whenever possible and military force when necessary to control the ongoing unrest. His Philippines' service earned him the Distinguished Service Cross.

The punitive expedition to Mexico

In January 1914, Pershing assumed command of an infantry brigade at the Presidio of San Francisco. Relations with Mexico had gravely deteriorated and by 1913 America's southern neighbour was beset by revolution and anarchy. In January 1914, his brigade was re-assigned to Fort Bliss, Texas, where it assumed responsibility for the security of the US–Mexican border.

Disaster struck in August 1915 when Pershing's family quarters at the Presidio were destroyed by a raging fire that took the lives of his wife and his three daughters. Only Warren, his 6-year-old son, survived. Those closest to him believed Pershing never recovered from the anguish of losing his beloved wife and daughters in such a manner.

In March 1916, the Mexican rebel Francisco 'Pancho' Villa raided the border town of Columbus, New Mexico, killing a number of American soldiers and civilians and triggering an immediate response from President Woodrow Wilson. Pershing was ordered to form a 10,000-man punitive expedition to Mexico to track down and capture Villa.

The expedition penetrated deep into the wastelands of northern Mexico and, although the force engaged in minor skirmishes, the search for Villa proved fruitless and the rebel leader was never captured. The expedition was ill-prepared and badly equipped for an extended foray into the Mexican wilderness and received little cooperation from the Mexican government, while the US government placed restrictions on Pershing; it ended in failure in January 1917. The only event of note was that Julio Cárdenas, Villa's chief bodyguard, was killed in a shoot-out by one of Pershing's aides, a young cavalry officer named George S. Patton.

The First World War

The United States entered the war in April 1917, with the promise of an expeditionary force, as an ally of Britain and France. The decision to create an American Expeditionary Force (AEF) to send to France required the ablest officer to be its commander-in-chief. When the only other candidate, General Frederick Funston, died in February 1917, Major General John J. Pershing was appointed by Wilson and promoted to the rank of four-star general.

The new AEF commander was given virtual carte blanche by Wilson and Secretary of War Newton D. Baker. The task of creating and training an army for war was one of the most daunting ever faced by an American military commander. A force had to be formed, armed and trained for battle. The US army of 1917 was rife with bureaucracy and duplication, woefully untrained and barely exceeded 100,000 men. It lacked even the rudimentary tools of war, possessing a mere 285,000 Springfield rifles, just over 500 field guns and only enough ammunition for a nine-hour bombardment. Pershing carried out his mandate with ruthless efficiency, cutting through red tape to create a force that by war's end numbered over 2 million men.

When he arrived in Paris in the summer of 1917, huge crowds greeted Pershing with near-euphoria as a conquering hero. France was at a low point of the war; its men were being squandered in futile battles for a few yards of useless terrain at places like the Somme, Ypres and Verdun, which became ghastly monuments to the folly of trench warfare.

Pershing came under unrelenting pressure from the French government and French and British generals to place American forces under their command, but steadfastly refused to bend. The AEF, he declared, would remain independent and would only fight under American command. Pershing's legendary resolve prevailed despite attempts in 1918 by the Allied Supreme Commander, French Marshal Ferdinand Foch, to marginalize the American role in the war. 'While our army will fight wherever you may decide, it will not fight except as an independent American army,' he told Foch. The French president Clemenceau once called Pershing 'the stubbornest man I ever met'.

However, a larger problem loomed. Before the AEF could fight it had to be trained for battle. The troops arriving in France were raw recruits. Having seen for himself the appalling conditions at the front and a French army in such disarray that it had mutinied, in October 1917 Pershing announced, 'The standards for the American army will be those of West Point', a worthy ideal but one impossible to obtain with a largely conscript army. Nevertheless, he was determined to make the AEF the best fighting force in the war. Few escaped his stern presence; one of the few who did was a captain named George C. Marshall, who daringly challenged Pershing's criticism of the 1st Division. Most thought Marshall's career over, but instead Pershing promoted him to play a key role in the AEF.

Pershing established his headquarters far from Paris, at Chaumont, instituted a series of battle schools to train officers and men in every aspect of warfare, and tapped Patton to form and train an American tank force that fought for the first time in the Battle of Saint-Mihiel in September 1918.

The AEF soon came to bear Pershing's unmistakable stamp and by the autumn of 1918 consisted of two field armies. Pershing and the AEF won the admiration of its once-sceptical allies. Although his generalship in France was not without its flaws, it is to his eternal credit that he refused to permit American soldiers to become the same cannon fodder in the trenches as those of his British and French allies.

In recognition of his achievements in the war Pershing was promoted in September 1919 to General of the Armies, the highest rank ever accorded an American soldier. His popularity was so high that he was considered a possible Republican candidate for president in 1920, but Pershing declined to campaign and the nomination instead went to Warren G. Harding, who won the presidency.

Later years and legacy

Pershing served as chief of staff of the army from 1921 until he retired from active service in 1924 at the age of 64, a revered figure in both the nation and as the army's grand old man. Less well known is his service from 1923 to the time of his death as chairman of the American Battle Monuments Commission (ABMC), created that year to oversee the operation and maintenance of American military cemeteries overseas. One of the junior officers who served under Pershing in the ABMC was Major Dwight Eisenhower, who authored the manual on American memorials in France.

Pershing's years of retirement were private and unremarkable. He wrote his memoirs, which won the Pulitzer prize for history in 1932, and maintained his stature as America's most senior and respected general. In later life his health deteriorated and in the four years prior to his death he was a patient at Walter Reed Army Hospital, where a regular stream of visitors came to pay their respects.

'Black Jack' Pershing died in Washington on 15 July 1948 and was buried in Arlington National Cemetery near the men of the AEF he had commanded. Despite his elevated rank, Pershing's plain white cross is the same as that of every soldier, sailor and airmen. Basil Liddell Hart later said of Pershing: 'There was perhaps no other man who would, or could, have built the American army on the scale he planned, and without that army the war could hardly have been saved and could not have been won.' Charles Dawes, his close friend and chief of supply in the AEF, and later vice-president under Calvin Coolidge, may have summed up the plain-spoken Pershing's legacy when he noted that: 'John Pershing, like Lincoln, recognized no superior on the face of the earth.'

His reputation as one of America's foremost and most

resolute soldiers endures. As his official biographer notes: 'For two-score years he had soldiered; his career shaped the army and sustained his nation.' To this day, John J. Pershing remains a symbol of perseverance and iron will.

KEMAL ATATÜRK

1881–1938

IAN BECKETT

STATUES OF MUSTAFA KEMAL, known since 1934 as Atatürk ('Father of the Turks'), can be seen everywhere in modern Turkey. Inevitably, there are mythic elements in the heroic version of his life, but there is no doubt that, out of the disintegration of the Ottoman Empire, he forged a Westernized and secular state that has endured for over eighty years. That he was able to do so rested on a military reputation first made at Gallipoli in April 1915, and upon his subsequent victories in the Greco-Turkish War of 1920–22.

Atatürk was born some time in 1881 in the Macedonian city of Salonika (now Thessaloniki in northeast Greece, then under Turkish rule). Initially he was simply called Mustafa, but at his military secondary school he was given the name Kemal ('Perfect') to distinguish him from other Mustafas. Despite opposition from his mother, he settled on a military career at an early age, passing through the War College to graduate as a lieutenant in 1902. Although of medium height and light build, Atatürk was nonetheless an imposing figure, and, with his fair hair, blue-green eyes and light complexion, by no means obviously Turkish: the sultan once referred to him as a 'Macedonian revolutionary of unknown origin', though it is more likely that he had Albanian antecedents. Atatürk was undoubtedly intelligent and shrewd, though also vain, cynical, domineering and unscrupulous. He drank heavily, played poker incessantly and had a prodigious sexual appetite.

The Young Turks

Atatürk's father, who died when Atatürk was 7 or 8, had been liberal-minded, and Atatürk was also influenced by the progressive liberal nationalism then popular among younger army officers. Indeed, he was arrested shortly after graduation from the Staff College in 1905 for being in possession of banned books – although, because of his youth, he was simply posted to a remote region in Syria. Atatürk was involved on the peripheries of subsequent military conspiracies that led to a coup staged by the army-based Committee of Union and Progress (CUP), popularly know as the Young Turks, in July 1908. This forced Sultan Abdul

Hamid II to restore the liberal constitution that had been suspended in 1877. However, Atatürk was not directly involved in the suppression of a counter-revolution by the Young Turks in April 1909, after which they forced the sultan's abdication in favour of his younger brother, Mehmet V. Nor did he play a role in the coup orchestrated by the minister of war, Mehmet Enver Pasha, in January 1913, by which the Young Turks seized total power.

The final straw for Enver and his colleagues had been Turkey's humiliation in the Italo-Turkish War of 1911–12 and the First Balkan War of 1912. Atatürk himself served against the Italians in Libya and in the Second Balkan War in 1913. The Young Turks had ambitions to revive Ottoman power and overthrow the settlement reached at the conclusion of the Balkan Wars. The remaining Turkish frontiers in Europe were now vulnerable to the territorial ambitions of Greece and Bulgaria, while Russia clearly retained its interest in the Dardanelles and the Bosphorus. Some Young Turks, therefore, favoured alliance with one of the great powers, while others wished to reach an accommodation with all the powers as a means of securing Turkey's future. From Enver's perspective, Germany had no apparent territorial ambitions on Turkish territory. By contrast, Britain and France seemed indifferent to Turkish sensitivities and were aligned with Russia.

Accordingly, with war breaking out in Europe, on 2

August 1914 Enver secured a secret treaty of alliance with Germany. Anxious to bring Turkey into the war, the Germans accepted Enver's invitation to give safe refuge in Turkish waters to two warships, *Goeben* and *Breslau*, which were being pursued by the Royal Navy in the Mediterranean. Their passage of the Dardanelles straits on 10 August violated international law, though the Turks claimed they had purchased the ships. Enver increased pressure on his colleagues by securing a German financial loan and then, on 29 October, by authorizing *Goeben* and *Breslau* to commence raids on Russian shipping and coastal towns on the Black Sea. Russia declared war on Turkey on 2 November, followed by Britain and France on 5 November; the Turkish counter-declaration was issued on 11 November 1914.

The First World War

The Dardanelles were the most obvious place for Britain and France to attack in support of Russia and, in February 1915, Atatürk, though still only a lieutenant colonel, was appointed to command the Turkish 19th Division on the Gallipoli peninsula guarding the north side of the straits. A British and French naval operation to force the straits began on 19 February but, when it faltered on 18 March, the Allies resolved to land at Gallipoli. It so happened that Atatürk had planned an anti-invasion exercise for his command on the very day, 25 April 1915, on which the British and French landings commenced.

The initial landings of General Sir Ian Hamilton's Mediterranean Expeditionary Force on the Gallipoli peninsula took place on 25 April 1915 at five beaches at Cape Helles in the south and, 12 miles to the northwest, at what became known as Anzac Cove. It had been intended to land the Anzacs (the men of the Australian and New Zealand Army Corps) adjacent to the Gaba Tepe headland. In the pre-dawn darkness, however, the current swept the boats to a narrow beach under steep cliffs about a mile and a half further north, between 'Hell Spit' and Ari Burnu.

There was little initial resistance, but confusion and the difficult terrain meant that it proved all but impossible to move inland quickly, though small parties began to climb the cliffs. Atatürk's division held the high ground at Chunuk Bair and Sari Bair. Getting no answer when he reported the landing to his superiors, Atatürk ordered an immediate counter-attack on his own initiative. Rushing to the front with just one junior officer, he found some of his men withdrawing. When they said they had no ammunition, he ordered them to use the bayonet instead, and by the sheer force of his personality forced them to take up new positions. As more of his force arrived so he pushed them forward, famously saying, 'I do not order you to attack, I order you to die. By the time we die, we will be replaced by other troops and commanders.'

Atatürk's vigorous counter-attack amid the scrub-covered gullies and valleys that characterized the terrain behind the cliffs prevented any further exploitation by the Anzacs. Indeed, men began to struggle back down to the beach and to take shelter there, and most of the artillery landed was taken off again as the situation deteriorated. By midnight the position seemed so critical that the Anzac corps commander, Sir William Birdwood, and his two divisional commanders, Major-Generals Bridges and Godfrey, urged withdrawal. They were overruled by Hamilton, who ordered them to 'Dig, dig, dig until you are safe.'

The campaign now effectively became a struggle for command of the high ground. From the Allied perspective, too much was asked of troops with inadequate artillery support, arising not so much from a lack of guns but from the technical limitations of artillery faced with the steep elevations and with few means of accurately locating Turkish guns. The Turkish defenders, too, suffered from the technical limitations of artillery, and had relatively few machine-guns, but Atatürk continued to inspire the defence, reminding his command in May 1915, 'Every soldier struggling here along with me should know that to carry the honourable task given to us not one step backwards will be taken.' His counter-attacks were repulsed with heavy losses, but he always had sufficient forces and supplies to reproduce the stalemate already encountered on the Western Front.

Atatürk's hastily improvised defence opposite the Australian and New Zealand Army Corps at 'Anzac Cove' halted their advance inland to the Sari Bair heights. Atatürk led another successful defence when fresh British forces landed at Suvla Bay on 6 August, by which time he was commanding the Turkish XVI Corps. At one point, he was hit by a shrapnel fragment, but the watch in his breast pocket, which was shattered in the impact, saved his life. Atatürk contributed significantly to the Allied defeat at Gallipoli, although the repeated massed counter-attacks he ordered were extremely costly.

Following the final evacuation of the Allies from Gallipoli in January 1916, Atatürk served in the Caucasus, Syria and Arabia, ending the war in command of an army group in Syria, though still only a brigadier general. Atatürk had become increasingly critical of the CUP leadership, particularly Enver, who he believed had blocked his promotion. Enver, however, fled Constantinople as British forces entered Damascus and Aleppo in October 1918 and advanced towards Mosul. An armistice was concluded on 30 October 1918, the Allies occupying the Dardanelles and the Bosphorus.

The revolt against the sultan

Sultan Mehmet VI, who had succeeded his brother in July 1918, received the peace terms as the Treaty of Sèvres on 10 August 1920. They reflected Britain and France's wish to preserve some kind of Turkish state, though dividing much of the Ottoman Empire between them and giving independence to Armenia and autonomy to the Kurds. Controversially, the treaty rewarded the Greek prime minister, Eleutherios Venizelos, for his wartime support of the Allies by awarding Greece most of the Aegean islands, the Gallipoli peninsula, Smyrna (Izmir) and Thrace. Indeed, in May 1919, with the Italians absenting themselves from the Paris Peace Conference during a dispute over their own irredentist claims in the Adriatic, Venizelos was invited to occupy Smyrna in order to forestall any Italian intervention, the Italians having landed troops at Adalia (Antalya) in March 1919. The Greek landing at Smyrna on 15 May 1919 was the catalyst for Atatürk's nationalist revolt.

Between December 1918 and March 1919 a national resistance movement emerged in the Turkish army, which had been drawn back into Anatolia at the armistice. Atatürk was inspector general of the Ninth (later renamed the Third) Army with the task of restoring order on the Black Sea coast when the Greeks landed at Smyrna. Ostensibly, his appointment was to reassure the Allies that there would be no massacres of Greeks or other Christians. In reality, Atatürk's appointment was always intended to further the establishment of the new national movement away from Constantinople. It was the first time he had secured a position of real authority, and he began contacting army commanders and provincial governors. On 22 June 1919 he issued a declaration at Amasya, calling on Turks to resist foreign domination and to forge a new central national body free of outside influence and control. In the declaration he also summoned a national congress at Sivas, although this was pre-empted by a proposed national congress of the Defence of Rights Associations at Erzurum. Atatürk agreed to attend, resigning from the army on 8 July 1919 when his intended presence led the sultan to order his recall and arrest.

All the principal army commanders had agreed to follow Atatürk's orders, and at the Erzurum and Sivas congresses the national movement effectively pledged itself to maintain the integrity of the Turkish state and to elect a national assembly to mandate a government capable of negotiating a satisfactory peace agreement with the Allied powers. Following negotiations with the government, elections were held and a new assembly convened in Constantinople in January 1920. Though this assembly passed the National Pact on 28 January, by which it was agreed to seek an independent Turkish state within secure frontiers, the deputies were not prepared to consent to the formation of the single national party that Atatürk had asked for. In the event, a serious clash between Turkish and French troops in Cilicia on the borders of Turkey and Syria, and the accompanying slaughter of an estimated 20,000 Christians, led to the Allies occupying Constantinople on 16 March 1920. Then, with Allied backing, the sultan established a gendarmerie called the Army of the Caliphate to suppress the nationalists. Atatürk was condemned to death in absentia for treason. His response was to summon a national assembly in Ankara on 23 April, which elected him provisional head of state by a single vote. In June the sultan accepted the Treaty of Sèvres, at the very moment when the Turkish army was beginning to prevail against the untrained gendarmerie.

The British and French had only limited forces available. Consequently, they responded to Atatürk's advance by asking the Greeks to launch offensives in Anatolia and Thrace. Throwing some 150,000 men into the offensives, the Greeks enjoyed considerable success, since Atatürk had at best 70,000 men available. At this point, on 2 October 1920, King Alexander of Greece was bitten on the leg by one of his pet monkeys, which he was trying to separate from his spaniel. Blood poisoning set in, and the king died on 25 October.

Alexander's young wife was expecting a child, but it

had been a morganatic marriage. Venizelos tried to get Alexander's younger brother Paul to accept the throne, but he refused unless the Greek people rejected both his father, the exiled King Constantine, and his older brother, former Crown Prince George. Opposed to Greek entry into the war and considered pro-German, Constantine had in 1917 been forced by Allied pressure to abdicate in favour of Alexander. Now, in the face of Prince Paul's conditions, Venizelos consented to an election in the belief that he would win it, but lost. On 19 December 1920 Constantine duly returned to the throne; incensed by this unexpected development, the Allies cut off all financial and military assistance, and withdrew their political support for the Greek action in Anatolia. Though he had opposed Greece entering the war against Turkey, Constantine decided not only to follow the Venizelist policy but also to step up the scale of a war that was relatively popular with the Greek public.

The Greco-Turkish War

Capitalizing on his good fortune, Atatürk appeared conciliatory towards Britain and France, while also reaching an accommodation with the Bolsheviks in Russia. Faced with problems in Syria and Lebanon, the French reached a secret understanding with Atatürk in October 1920 and withdrew from Cilicia; by June 1921 the Italians had also withdrawn their small force from southwest Anatolia. Meanwhile, Atatürk's army gave ground to the Greeks at Eskisehir in January 1921, but checked the Greek advance in the First and Second Battles of Inönü (6–10 January and 23 March – 1 April respectively), largely under the direction of Atatürk's chief of staff, Ismet Pasha. Now under Constantine's personal command, the Greeks advanced again, and in July 1921 forced the Turks back to the River Sakarya, only 50 miles from Ankara. Atatürk was much criticized for authorizing the retreat, but it was undoubtedly the right decision. Launching what they hoped to be a decisive offensive, the Greeks advanced again across the arid steppe of central Anatolia, but in a grim, attritional 22-day battle on the Sakarya, from 23 August to 13 September 1921, the Greeks were forced to retire with their supplies exhausted.

The initiative was now in the hands of Atatürk, who, elevated to field marshal and *Ghazni* ('warrior for the faith'), had become commander-in-chief in August 1921. Atatürk implemented total mobilization, putting some 78,000 men into the field, and on 26 August 1922 he launched his own offensive, feinting in the north and attacking in the south at Afyon Karahisar and Dumlupinar. Caught by surprise, the Greek army collapsed. The Turks re-entered Smyrna on 9 September 1922, as the Allies evacuated over 213,000 people from the burning city.

With the Greeks now expelled, there remained the British occupation forces at Chanak on the Dardanelles. After Turkish forces appeared opposite them on 23 September 1922, Lloyd George's government instructed the British commander at Constantinople, General Sir Charles 'Tim' Harington, to issue an ultimatum demanding an immediate Turkish withdrawal, while at the same time also seeking the support of the other Allied powers and the dominions. The French and the Italians had no intention of becoming entangled in a war with the Turks and, among the dominions, only New Zealand and Newfoundland were prepared to stand by Britain. In any case, Harington declined to pass on the ultimatum to Atatürk, stressing the likely dangers for the Christian population in Constantinople if he had to draw his forces out of the city as immediate reinforcements for Chanak. Harington also believed Atatürk had no real wish to fight and would negotiate. Indeed, Harington, who had already made tentative approaches to Atatürk, negotiated an armistice with Ismet on 11 October 1922 by which it was agreed to withdraw all remaining Allied and Greek forces from Turkish soil once a new peace treaty had been concluded. The conference began at Lausanne on 20 November. In the meantime, Lloyd George's coalition government had collapsed.

'Father of the Turks'

Atatürk's victories enabled him to move to abolish the sultanate on 1 November 1922, and less than a week later Mehmet VI had taken refuge on a British warship. The new treaty signed at Lausanne on 24 July 1923 revised the Sèvres terms to Turkey's advantage, with the restoration of Armenia, eastern Thrace, Smyrna and all of Anatolia. In the short term, the Treaty of Lausanne led to a mass exchange of Greek and Turkish populations; in the longer term, as the only peace treaty after the Great War that was negotiated rather than imposed, Lausanne was the only one to endure. The Dardanelles straits remained demilitarized, but were restored to Turkey in 1936. Ankara became the capital on 9 October 1923 and Atatürk became president of the new republic on 29 October 1923. Religion and state were then separated by the abolition of the caliphate in March 1924.

Thus, Atatürk was poised to effect what he had spoken of in June 1918 as a revolution in social life but in many ways the programme he now embarked on – to secularize, modernize and Westernize Turkey, with the aim not of imitating the West but of participating with it as an equal – merely completed the liberal reform programme of the Young Turks. In the event, secularism has often proved divisive, and there was no room in Atatürk's Turkey for ethnic diversity. Nonetheless, it was not inappropriate that

Mustafa Kemal should assume the name of Atatürk in November 1934, for he had certainly secured Turkey's continued independence. Characteristically perhaps for a larger-than-life figure, he died in November 1938 of cirrhosis of the liver.

BASIL LIDDELL HART

1895–1970

ANDREW ROBERTS

'THE CAPTAIN WHO TAUGHT GENERALS' was the soubriquet commonly used of Sir Basil Liddell Hart, the most famous and influential military historian and journalist of the twentieth century. There are others who can lay claim to have invented the ideas behind mechanized warfare between the two World Wars, but it was Captain Liddell Hart who popularized, politicized and propagandized these vital concepts. Since his theories of armoured warfare and the 'expanding torrent' were to play important roles during the Second World War and beyond, Liddell Hart deserves his place alongside Clausewitz in this book although, like the celebrated Prussian, he was more a great teacher than a great commander.

Born in 1895, the son of a Wesleyan Methodist clergyman based in Paris, Basil Liddell Hart was descended from

Gloucestershire yeoman farmers. Despite his delicate health he formed a fascination for warfare from childhood; when playing with toy soldiers he would manoeuvre them strategically rather than simply knocking them down as other children might. He attended St Paul's School from 1911 to 1913, and then Corpus Christi, Cambridge, where he read History without much distinction. Despite having bad enough eyesight to exempt him from military service in the Great War if he so chose, he begged his parents' permission to join up and was commissioned into the King's Own Yorkshire Light Infantry (KOYLI) in December 1914, aged only 19.

Experience of the trenches

As well as fighting at Ypres in November 1915, Liddell Hart saw action on the first day of the Somme offensive, escaping unscathed because he was placed in the reserve. No fewer than 450 men in his 800-strong battalion were lost between 1 and 3 July 1916 in that engagement. 'All the KOYLI have suffered badly,' he wrote home, 'two other service battalions having lost all their officers without exception, and nearly all their men. I have never lost so many friends before, all my friends in the various battalions which I know having been wiped out.' After a week of recuperation he was sent back to the line for a renewed offensive.

Between 16 and 18 July 1916, Liddell Hart disappeared in Mametz Wood, the scene of some of the most vicious fighting of the entire war. It is unclear to this day precisely what happened to him, and there are some slight indications from his writings that he might have suffered from a panic attack there, but when he emerged he was certainly suffering from the effects of phosgene gas. 'He abhorred war,' writes his biographer Alex Danchev. 'He abhorred its irrationality, its lumpishness, its contagion, its waste.' Liddell Hart himself described it as 'a farcical futility'.

The evangelist of the tank

The gas attack left Liddell Hart with a 'disordered action of the heart', but despite that he stayed in the army until 1924, before finally leaving to become assistant military correspondent of the *Morning Post,* which later became the *Daily Telegraph*. The following year he became its chief military correspondent and later stated: 'I decided to make it a platform for the mechanization of the army.' He stayed at the *Telegraph* until 1935, becoming easily the most influential military commentator in the history of journalism.

Of course in a sense armies had been mechanizing ever since the invention of the machine-gun in the mid nineteenth century, a weapon that by 1914 had made cavalry obsolete. The best way to attack enemy command and control posts across country was the constant concern of military theorists of the inter-war years, such as generals J. F. C. Fuller, Giffard Martel, Percy Hobart as well as Liddell Hart, and their unanimous answer was the tank, which had, after all, been used in warfare ever since its first effective use during the Battle of Cambrai in 1917. Nonetheless this was still highly controversial since in the early 1920s tanks still had no radio communications, were incredibly slow, often broke down, had limited firepower and relatively little cross-country capability.

'Soldiers are sentimentalists,' wrote Liddell Hart, 'not scientists.' His relationship with the War Office started off well, and he received leaks from the pro-mechanization lobby on the Army Council, invitations to manoeuvres, letters from successive chiefs of the Imperial General Staff and occasionally classified documents were passed to him if they put the army in a good light. Yet by the late 1920s it was clear to the High Command that he put his journalistic career before his popularity with them, and his harsh criticisms of their perceived military conservatism began to rankle.

As a Somme veteran, Liddell Hart had the necessary *locus standi* in inter-war Britain to seem to speak for those who had lost their family, their limbs or their youth to what he denounced as 'progressive butchery, politely called attrition', the strategy he accused Haig and others of having pursued on the Western Front. He wrote no fewer than thirty books expounding his theories, occasionally falling foul of the (hardly cardinal) sin of self-plagiarism and, as Danchev states with commendable understatement: 'He did not suffer from modesty.'

Liddell Hart and the Second World War

The outbreak of the Second World War gave Liddell Hart the opportunity to say 'I told you so' on a very regular basis for six years. As the very method of air-supported, tank-led, highly mobile warfare that he had been advocating for two decades – blitzkrieg – flashed over Poland, Norway, France, Belgium and Holland like the lightning it was named after, his theories were proved right.

The only problem was that it was the Germans rather than the Allies who were putting them into ostentatiously good practice. At 4 p.m. on 9 March 1943, General Heinz Guderian – along with Rommel the apostle of blitzkrieg – addressed Hitler and virtually the entire German Army High Command (OKW) in what developed into a hard-fought, four-hour meeting. In order to ram home his points about the capabilities of highly mobile mechanized forces on the Eastern Front, Guderian read out an article by Liddell Hart on the organization of armed forces, past and future.

The following year, 1944, Liddell Hart deduced the exact timing and place of the D-Day landings, not, as MI5 feared, because of leaks from his army contacts, but from a deep study of the geography and tidal flows of north-west France. In the latter part of the war he opposed the Allied area-bombing strategy against Germany, and in 1945 denounced the use of the atomic bombs against Japan. After the war he campaigned against nuclear weapons, and shortly before his death in 1970 described himself as a pacifist.

The strategy of the 'expanding torrent'

In November 1920, only two years after the armistice that ended the Great War, Liddell Hart expounded his 'Strategy of the Expanding Torrent' to an audience at the Royal United Services Institute (RUSI), by which an army 'would ensure that the momentum of the attack was maintained right through the whole of the enemy's system of defence, which might be miles deep'. By the following year he accepted that, to achieve this, the tank was the weapon of the future. He was persuaded in this by his mentor 'Boney' Fuller, its first major advocate, with whom Liddell Hart was to have a strange intellectual love–hate relationship over the coming decades.

'The development of mechanical firepower has negatived the hitting power of cavalry against a properly equipped enemy,' Liddell Hart was to write in his 1927 work *The Remaking of Modern Armies*. 'But on land the armoured caterpillar car or light tank appears the natural heir of the Mongol horseman, for the "caterpillars" are essentially mechanical cavalry. Reflection suggests that we might well regain the Mongol mobility and offensive power by reverting to the simplicity of a single highly mobile arm.' Although Liddell Hart probably never actually sat in a moving tank, he spotted their potential early on and hailed every advance in firepower and motorization that made them more efficient.

In retrospect it seems incredible that there was so much official scepticism over the likely future combat role of the tank. Yet in the 1920s, mechanization was considered 'that fearful fate which hung like a shadow of doom over all cavalry regiments at that period'. Today it is hard to see how anyone could have held out against full-scale mechanization, considering how tanks had been used to good effect at the Battle of Cambrai in 1917 and since, yet it was so.

As late as 1936 Alfred Duff Cooper, then Secretary of State for War, admitted to the eight cavalry regiments that were about to be mechanized: 'It is like asking a great musical performer to throw away his violin and devote himself in future to the gramophone.' Cooper's sole foray into fiction, the superb 1950 novel *Operation Heartbreak*, features an argument between two British officers on the prospect of mechanization a quarter of a century earlier:

> 'I'd as soon be a chauffeur,' exclaimed Willie passionately one evening, 'as have to drive a dirty tank about and dress like a navvy.' 'Of course,' replied Hamilton blandly, 'if all you care about is wearing fancy dress, playing games on horseback, and occasionally showing off at the Military Tournament, you're perfectly right to take that view; but if you're interested in war, or even hoped to take part in one, you'd be praying that your regiment might be mechanised before the next war comes.'

In the War Office and Westminster there were those who for an inordinately long time agreed with Rudyard Kipling, who had described the 1930 Salisbury Plain mechanized manoeuvres as 'smelling like a garage and looking like a circus'. Liddell Hart recalled how when Percy Hobart gave a demonstration of tank warfare on Salisbury Plain four years later, 'orthodox soldiers retorted that such a method would not work in war'. The New Zealand cartoonist David Low even claimed to have overheard one cavalry officer telling another in a Turkish bath in the 1920s that if mechanization came, cavalry uniforms should nonetheless stay the same, right down to the spurs. (That moment proved the inspiration for his cartoon character Colonel Blimp.)

Yet the tank offered a way out of the slow stalemate

of the trenches. For one who had fought at Ypres and the Somme, it was unsurprising that Liddell Hart considered speed as of the essence in any future conflict. 'Of all qualities in war it is speed which is dominant,' he wrote in a book that arose from lectures he had delivered at RUSI in 1922,

> speed both of mind and movement, without which hitting-power is valueless and with which it is multiplied, as the greatest of all commanders [*i.e. Napoleon*] realised in his dictum that force in war is mass, or as we should better interpret it under modern conditions, firepower, multiplied by speed. This speed, only to be obtained by the full development of scientific inventions, will transform the battlefields of the future from squalid labyrinths into arenas wherein manoeuvre, the essence of surprise, will reign again.

His hatred of trench warfare, which he called, with typical regard for alliteration, 'mausoleums of mud', was evident.

Along with the tank, Liddell Hart was an exponent of air power, as he had been ever since seeing a Zeppelin raid on Hull in 1915. 'One could see the gleam of light each time a trapdoor opened to drop a bomb,' he later recalled. To coordinate air power – specifically the dive-bomber –

with tank and infantry support would, so Liddell Hart preached, be to effect the expanding torrent theory on a future battlefield, and this would be the key to victory.

The strategy of the indirect approach

What Liddell Hart called the 'Strategy of the Indirect Approach' was developed in 1928–9, and was followed up by his related book *The British Way in Warfare* three years later. It effectively argued that British grand strategy was best served not by huge direct continental commitments – such as he had experienced on the Somme – but by wearing the enemy down by blockade, bombing and attacks on the periphery. (Less convincingly, Liddell Hart also believed that his theory could be applied to practical philosophy, religious disputation, the art of salesmanship and was even 'fundamental to sex life'.)

'Throughout the ages,' wrote Liddell Hart, 'effective results in war have rarely been attained unless the approach has had such indirectness as to ensure the opponent's unreadiness to meet it. The indirectness has usually been physical, and always psychological.' This was a paean to 'the art of outflanking' (which was the title his friend Robert Graves suggested for it). 'In strategy,' summarized Liddell Hart, 'the longest way round is often the shortest way home.'

Danchev believes that the book *Strategy: The Indirect Approach* (1941) was 'as near as Liddell Hart ever got to a treatise, an *essai général*, of his own'. Liddell Hart cited Cromwell at the battles of Dunbar and Worcester, and Marlborough's intensive marches and counter-marches, as examples of British attempts to turn flanks and, if possible, cut the enemy's lines of communication and retreat.

The British way

'This much is certain,' Liddell Hart said at RUSI in 1931, unveiling his 'British Way in Warfare' concept, 'he that commands the sea is at great liberty, and may take as much or as little of the war as he will. Whereas those that be strongest by land are many times nevertheless in dire straits.' Britain should, he argued, fight by exercising naval power, financing allies ('lending sovereigns to sovereigns'), shoring up trade, making minor amphibious landings, but essentially letting other powers fight the huge land engagements.

Yet as Sir Michael Howard has pointed out, such a strategy was usually forced on Britain through *force majeure*, and while it might have allowed Britain to survive, 'it never enabled us to win'. What allowed Britain to win historically, Howard argued convincingly, was instead 'a commitment of support to a Continental ally in the nearest available theatre, on the largest scale that contemporary resources could afford'. Blenheim, Waterloo and D-Day were all fought under those circumstances. Even Winston Churchill and Field Marshal Sir Alan Brooke, who both followed the 'British Way in Warfare' as long as they could during the 'Indirect Approach' part of the Second World War between 1939 and June 1944, nonetheless had to accept that Hitler could not be defeated except by landing on the Continent a British army 'on the largest scale that contemporary resources could afford'. There was indeed a British way of warfare, but it was not the one that Liddell Hart prescribed, and it was not based on the indirect approach, at least not in the last analysis.

Yet today the philosophy of the Indirect Approach – principally, of course, through the medium of international terrorism – is suddenly central to modern warfare, in ways Liddell Hart could not have predicted in 1941. As Danchev pointed out in 1998, 'Contemporary military doctrine is suffused with the indirect approach'. And that was even before 9/11.

Liddell Hart on Blitzkrieg

On reading the German General Heinz Guderian's account of the 1940 blitzkrieg campaign, Liddell Hart wrote of how it almost felt as if one were riding in a tank beside the general himself, and: 'For me it was like the repetition of a dream, as it was just the way that in pre-war years I had pictured such a force by a leader who grasped the new idea – only to be told, then, that the picture was unbelievable.'

In his 1950 autobiography, *Panzer Leader*, Guderian was unambiguous:

> It was Liddell Hart who emphasized the use of armoured forces for long-range strokes, operations against the opposing army's communications, and also proposed a type of armoured division combining panzer and panzer-infantry units. Deeply impressed by these ideas I tried to develop them in a sense practicable for our own army. So I owe many suggestions of our further development to Captain Liddell Hart.

Guderian later told Liddell Hart he had first read his articles in 1923–4, and that in the 1924 manoeuvres, he had been told by the inspector regarding tanks: 'To hell with combat! They're supposed to carry flour!'

Liddell Hart wrote a Foreword to Guderian's book. It was nominally about Guderian, but it could equally be read more autobiographically. Of blitzkrieg, Liddell Hart wrote of how the great Panzer commander shaped history 'by means of a new idea ... a new idea of which he was both the exponent and executant'. Guderian, he believed, personified 'the quintessence of the craftsman in the way he devoted himself to the progress of a technique ... To understand him one must be capable of understanding the passion of pure craftsmanship.'

Similarly, there might have been more than a trace of self-reference (and self-regard) in Liddell Hart's estimation of Guderian:

> Most of the recognized masters of the art of war have been content to use the familiar tools and techniques of their time. Only a few set out to provide themselves with new means and methods ... Developments in tactics have usually been due to some original military thinker and his gradually-spreading influence on progressive-minded officers of the rising generation.

Liddell Hart on the Great Commanders

In a book the title of which contains the words great commanders, it is worthwhile repeating Liddell Hart's estimation of what he called 'the qualities that distinguished the Great Captains of history'. These included:

> *coup d'œil,* a blend of acute observation and swift-sure intuition; the ability to create surprise

> and throw the opponent off balance; the speed of thought and action that allows an opponent no chance of recovery; the combination of strategic and tactical sense; the power to win the devotion of troops, and get the utmost out of them.

Because this checklist has not materially changed for centuries, and is unlikely to do so in the foreseeable future, the 'Captain Who Taught Generals' still has much to teach us.

CARL GUSTAF MANNERHEIM
1867–1951

ALLAN MALLINSON

MANNERHEIM'S GENIUS, like Napoleon's, lay in his grasp of the complete art of war. The great Finnish marshal practised that art, with conspicuous success, at every level, from the tactical to the grand strategic. He practised statecraft, indeed, and in a fashion that would have brought plaudits from Machiavelli and Clausewitz. In doing so he earned the respect of friends and enemies alike. And whereas the bitter northern snows were Napoleon's nemesis, 'General Winter' was Mannerheim's ally in his greatest trial: war with Stalin's Russia.

But while those snows seem still to bear the footprints of the Grande Armée – and later those of the Wehrmacht, which tried in vain to reach Moscow – they have largely covered those of the Finns, who fought in that early chapter of the Second World War, the 'Winter War' of 1939–40, and again in what became known as the 'Continuation War' of 1941–4. Mannerheim steered both Finland's armed forces, and the nation itself, to an outcome that none but a commander of the greatest integrity, foresight and skill could have achieved: the Finns fought the Russians to a standstill, twice, and with German help, yet in 1945 remained a free nation, unoccupied by the Red Army. Mannerheim is an exemplar – arguably one of the finest – of Clausewitz's concept of military genius, in particular its two indispensable components: 'first, an intellect that, even in the darkest hour, retains some glimmerings of the inner light that leads to truth; and second, the courage to follow this faint light wherever it may lead'.

Early years

Carl Gustaf Emil Mannerheim was born in Askainen, south-west Finland, on 4 June 1867, the third child of Count Carl Robert Mannerheim and his wife, Helena von Julin. The Mannerheim family's origins were Dutch, or possibly German, from the merchant families of Hamburg. The Marheins, as they were then called, emigrated to Sweden in the seventeenth century, and were later ennobled. Finland was a Swedish possession until 1809, when it became an autonomous grand duchy within the Russian empire. Like most of the Finnish nobility, the young Carl

Gustaf's first language was Swedish, and all his life he spoke Finnish imperfectly and with a pronounced 'foreign' accent.

His upbringing, though the family was by then in straitened circumstances, was idyllic: sledging, swimming, horses. Already exceptionally tall, at 13 he was sent to the Military Cadet School in the old fortress city of Hamina, on the southeast coast of Finland. Young Carl Gustaf neither enjoyed the school, nor did he excel, except at games. But he persevered, until one night, when he was 18, he decided that the cadet curfew was too restrictive, put a dummy in his bed, and slipped out of barracks. The offence brought instant dismissal.

It turned out to be the making of him, however. He polished his school-room Russian, and a year later was admitted to the Nikolaevsky Cavalry School in St Petersburg. Here he did well, and in 1889 was commissioned in the Alexandrijski Dragoons, quartered in Poland. But he knew that, for an ambitious Finn especially, he must get himself into a Guards regiment, which in 1891 he was able to do, joining the Chevalier Guards of the Empress in St Petersburg.

Regimental and court soldiering suited Mannerheim; he learned social and diplomatic skills, mastered several languages (by the end of his life, in addition to Swedish, Finnish and Russian, he spoke English, French, German, Portuguese and some Mandarin Chinese), and cut quite a dash. Contemporaries record that 'he was energetic, correct over money matters, a good sportsman, an excellent rider, and lived soberly'. Of the coronation procession of Tsar Nicholas II in 1896, one artist wrote, 'Gustaf Mannerheim walked before the emperor's baldaquin with drawn sabre and looked very handsome – truly imposing.'

Not surprisingly, he made a 'good' marriage – to Anastasia Arapova, the daughter of a former Guards general and an heiress. They had two daughters, but after ten years the marriage fell apart. For the rest of his life Mannerheim lived as a bachelor, later with some asceticism.

Active service

Guards soldiering in St Petersburg was agreeable, but it wasn't real soldiering. War with Japan, in 1904–5, gave Major Mannerheim his opportunity. 'I am 37 years old,' he wrote home. 'Serious campaigns do not often occur, and if I do not take part in this one, there is every chance that I shall become an armchair soldier, who will have to keep silent while more experienced comrades make the most of their wartime impressions.' Promoted lieutenant colonel of the Nezhin Dragoons in Manchuria, Mannerheim sought – and saw – a good deal of action, much of it chaotic. Rather like the young Arthur Wellesley in Flanders, he learned how not to do things. He, personally, distinguished himself, however, and was promoted colonel.

When the war ended the following year, the General Staff sent the promising young colonel of cavalry on a singular commission: a survey, on horseback, from Russian Turkestan to what was then Peking – a distance of almost 9,000 miles. It took two years, but his reports were well received. Colonel Mannerheim had proved himself a fine staff officer as well as a brave and capable commander. His star was in the ascendant: within five years he was promoted major-general and given command of the Emperor's Uhlans of the Guard in Warsaw.

The Great War

In August 1914 Major-General Mannerheim's brigade formed part of the scratch covering force for the mobilization of the Imperial Russian army. They were soon in action against the Austrians advancing into Galicia. At the Battle of Krasnik, 150 miles south of Warsaw, Mannerheim displayed such a mastery of initiative and manoeuvre that his brigade imposed a strategic check on the enemy – gaining four crucial days for the mobilizing army – for which he received his first decoration for gallantry, the Sword of St George. He was soon commanding a division.

Mannerheim's diaries and letters from this time show more than mere tactical flare, however. Rather, they reveal foresight and a rare grasp of the strategic and geo-political situation – not least an awareness of the ambivalence towards the war felt in Finland, where a pro-German sentiment had emerged out of opposition to Russian imperialism. Mannerheim was on leave in St Petersburg when in March 1917 the first shots of the Russian Revolution were fired, and had to flee his hotel in borrowed clothes – a useful taste of what might come in Helsinki if the revolutionaries seized power there.

In July Mannerheim was promoted lieutenant general to command VI Cavalry Corps in Romania, but the corps saw no action in the disastrous Russian offensive that summer. During sick leave in Odessa, he decided there was no future with any military integrity in Russian service, and set off back for Finland. Soon after, Lenin seized power.

The Bolshevik Revolution

'The White General' – as Mannerheim became known – managed, not without more close shaves, to get to Helsinki in December, where the new Finnish government at once made him commander-in-chief. Lenin had conceded independence to the Finns, under the terms of the Brest-Litovsk peace treaty with Germany, but in the expectation that the Finnish proletariat would overthrow the bourgeoisie and reunite with the new Soviet Russia.

Mannerheim had few forces with which to defeat the Finnish 'Red Guards' and eject the 40,000 Russian troops still in the country, most of them loyal to Lenin's provisional government. His masterstroke was in moving out of Helsinki, to Vaasa in the west, with its open lines of communication to Sweden. He had seen in Russia how a metropolitan senate and headquarters would inevitably be overthrown by the concentration of revolutionary forces, and that the fight back would then be against the odds. Away from Helsinki he would, at least, have balance.

At Vaasa he was able to build up an army based on the Civil Guards, and after three months of civil war, with the help of German troops, he defeated the Red forces. But it had been a fractious time, with Mannerheim at loggerheads with the pro-German elements of the government. He was exasperated, too, with Sweden, from whom little support materialized; indeed, Sweden's occupation of the Åland Islands was a lesson in the hazard of placing hope in Nordic solidarity.

After the conclusion of the civil war Mannerheim resigned, but when Germany finally collapsed in November 1918 he was recalled as regent. He stood in Finland's first presidential election the following summer, but was defeated by compromise voting among the centre parties, who were fearful of a split over his support for intervention in the Russian counter-revolution. He retired to private life, working for the Red Cross in Finland and for the League for Child Welfare, which he founded.

The gathering storm

In 1931 General Mannerheim was recalled once more – to be chairman of the new Defence Council, with the designation 'commander-in-chief in time of war'. Two years later he was made field marshal, and for the next five years he pressed for greater defence spending and rearmament. He was largely unsuccessful, but he did reorganize the army's mobilization, training and defence plans. Perhaps surprisingly for a cavalryman, he argued against the army high command's preference for a war of manoeuvre in the Karelian isthmus (the direct route from Leningrad to Helsinki), pressing instead for a line of fixed defences, the so-called Mannerheim Line.

Once again, he urged a Nordic mutual assistance pact, but Sweden would not be drawn. And when Stalin began pressing for border adjustments to strengthen Russia's naval defences in the eastern Baltic, Mannerheim advocated compromise: an exchange of territory, and leasing key islands. The government refused, but neither would it increase the defence budget. The field marshal was on the point of resigning when the Russians declared war, on 29 November 1939. At the age of 72, Mannerheim again took command of the country's forces, and moved to the headquarters at Mikkeli, from where he had directed the closing moves of the war of independence twenty years before. His first general order was of a Napoleonic grandeur and Churchillian insight:

> Brave soldiers of Finland!
> I enter on this task at a time when our hereditary enemy is once again attacking our country. Confidence in one's commander is the first condition for success. You know me, and I know you ...

No other nation's commander-in-chief at that time could have made so simple an assertion of mutual confidence: Mannerheim was an undisputed war hero, a general of proven ability, a patriot and a household name.

The Winter War

The Russians attacked through the Karelian isthmus, but with neither the strength nor the skill expected. After a precipitate withdrawal by the Finnish covering force – which Mannerheim checked by sheer force of personality – the army fought hard and effectively along the line of fixed defences. The Soviet advance ground to a halt, like that of the French before Wellington's Lines of Torres Vedras. But unlike Wellington's great defensive position, the Mannerheim Line was not anchored on impassable natural obstacles (the lakes were frozen): the field marshal knew the defences could be breached if the Soviets concentrated their force.

A second axis of advance north of Lake Ladoga fared no better at first. The Finns' mastery of movement in the snow-covered wilderness, and their *motti* tactics (encirclement and later destruction), inflicted huge casualties. But in January 1940 a new Russian commander, Timoshenko, was able to reopen the offensive on both axes with greater concentration of effort. With no practical help forthcoming from Sweden or the Allies, Mannerheim pressed the government to sue for peace.

Stalin was only too willing to negotiate: he had gained his immediate object – control of the isthmus and a zone of Karelia north of Lake Ladoga, safeguarding the strategically vital Leningrad–Murmansk railway. But it was at a prodigious cost in men, material and prestige – at least 120,000 dead. The cost to the Finns had been great, too: 27,000 dead, and, under the terms of the Moscow peace treaty, more territory than the Russians had originally demanded. But the moral effect of fighting the invader to a standstill was considerable. Mannerheim knew that although the country was now more vulnerable than before (the new Russian frontiers allowed little space to be traded in a covering action), paradoxically Finland's strategic position was improved.

Improved, but extraordinarily tricky. The peace was fundamentally unstable: half a million Finns had been displaced by the peace treaty, whose terms also allowed the rail transit of Russian troops. This would soon be paralleled – almost literally in places – by the rights given to

German troops to pass through Finland to occupied northern Norway. Finnish sovereignty was therefore highly fragile, and without help from the Allies or Sweden, the only source of military matériel and economic aid (grain especially, a critical factor in any future mobilization) was in supping with Hitler, if with the longest possible spoon. Germany was certainly keen to help.

Finnish motti tactics

Motti literally is a cubic metre of cut timber; traditionally in Finland, *motti* were cut and stacked throughout the forests for later collection. During the Winter War the Finns called encircled Soviet troops *Mottiryssä* or *Motti-Russki* – a gruesome analogy with firewood just waiting to be burned.

The Red Army's November 1939 offensive, in the wilderness of Karelia especially, was tied to narrow, single-track roads through snowbound forests. Unlike the Finns, the Russians were dependent on motor transport, and their rate of advance inevitably slowed; in places the long columns became bogged down and strung out over long distances, with small groups of vehicles becoming isolated, even more vulnerable to attack. The Russians were strangely uncomfortable in this environment – jumpy even: when the head of the advancing column was attacked, the rest of the vehicles would stop and adopt static defensive positions rather than attempting to counter-attack.

Once the column had been halted, Finnish infantry – ski troops especially – would begin a series of envelopments, and it was the enveloped forces that became known as *motti*. Initially the tactic was opportunistic, but one *motti*, at Suomussalmi, was fully planned from the outset, and on a much larger scale than hitherto: it succeeded in trapping the entire Soviet 44th and 163rd Divisions, cutting the line of supply and making numerous smaller *mottis* of the line of communication troops.

As a rule, the Finns attacked the weakest *motti* first, further isolating the stronger and less vulnerable pockets. Where the *motti* were too strong to overrun, Finnish ski troops, armed with sub-machine-guns, grenades, Molotov cocktails, satchel charges and smoke grenades, kept up pin-prick attacks, often infiltrating the defences, to harass the Russians and keep them off-balance. The severe winter did the rest. Thousands of Russian troops simply froze to death at their posts. Meanwhile the warmly clad and well-camouflaged Finns moved from one *motti* to another, attacking them at will and then disappearing into the forest. Soviet officers said they never actually saw a Finnish soldier, only their handiwork.

Attempts to break out of the *motti* were usually blocked successfully, for the Soviet troops were reluctant to abandon their vehicles. Re-supply by the Red Air Force was often their only salvation.

The unstable peace and the Continuation War

Through the rest of 1940 and the early months of 1941, Mannerheim prepared for war as best he could. When Hitler confirmed his intention to attack the Soviet Union, and his wish for a co-offensive with the Finns, Mannerheim recognized both a degree of inevitability (Stalin would regard them as de facto allies) and opportunity – perhaps the only opportunity – to re-establish secure borders. What followed in terms of campaign planning, with its Clausewitzian linkage to the ends of grand strategy, was really quite remarkable.

It all hinged on the fate of Leningrad. Mannerheim's object was the restoration of the 1939 borders, not the capture of Leningrad or German conquest of the Soviet Union. On the other hand, the German General Staff wished to tie down as many Soviet troops as possible in Finland. They calculated the Stavka (the Soviet High Command) would be prepared to withdraw troops north of Lake Ladoga for the defence of Leningrad, but not from the isthmus, for that would allow a rapid and easy advance to the city. Accordingly, OKW (the German High Command) pressed the Finns to mount their major offensive in the north. Mannerheim argued instead that he needed to concentrate his forces on the isthmus: it was there that the greatest threat lay, the 1940 frontier being so far west. He calculated that Stalin would recognize this as a move to restore the 1939 frontier rather than a move against Leningrad, which would not be the case with an offensive in the north.

Operation Barbarossa – the German attack on the Soviet Union – took Stalin wholly by surprise, despite Anglo-US warnings. The Finns were able to recover ground in the isthmus, but poor initial advances by the Germans allowed Mannerheim to launch a secondary offensive north of Lake Ladoga. He could now recover the 1939 border without the Soviets believing the Finns were part of a combined operation to take Leningrad. Indeed, the Finnish government was at pains to proclaim 'co-belligerency' and not alliance with Germany. Finnish troops advanced rapidly to the old frontier, but Mannerheim then took the decision to press on to the Svir River to secure a line of defence against the inevitable counter-offensive. The move was not without cost: Britain formally declared war on Finland on 6 December 1941 (a rare case of one democracy declaring war on another), though the British took only limited military action – principally air cover for the Red Army supplied by RAF Hurricanes based at Murmansk.

Fighting on the Finnish front eventually died down with the Soviets' need to strengthen the defences of Leningrad and beyond; but German reverses, especially Stalingrad, brought the anticipated counter-offensive, in June 1944. Mannerheim had calculated, however, that when the Western Allies opened up the second front, Stalin would withdraw troops from the northern front for the race to Berlin. His dilemma was how long to fight alongside the Germans before making peace. Hitler suspected this, and demanded a pact: no separate peace with Stalin, in return for continued, crucial, support.

Mannerheim and Risto Ryiti, the Finnish president, dealt with this by a desperate *ruse de guerre*. Ryiti sent Hitler a personal undertaking not to begin independent peace negotiations with Moscow. Military aid therefore continued, and the Red Army's advance was slowed. But after a month's fighting it was clear the pressure was irresistible. As planned, therefore, President Ryiti resigned, appointing Mannerheim in his place, who at once repudiated Ryiti's 'personal' undertaking to Hitler, and opened negotiations with Moscow.

The terms imposed by Stalin were severe: a return to 1940 borders, with other territorial appropriations and substantial reparations. They were also thorny: German troops were to be evicted. The alternative, however, was occupation. With this second treaty, President Mannerheim, Marshal of Finland (the rank had been created for him in 1942 for his 75th birthday), would at least achieve his grand strategic objective: an independent Finland.

The expulsion of Finland's former de facto ally was not without bloodshed, but it was achieved by a subtle use of force, ruse and the sheer power of Mannerheim's personality. And Finland was spared occupation by the Red Army, unlike the Baltic States and eastern Europe. Ill health forced Mannerheim's resignation in 1946, and he died five years later. But modern Finland is his legacy, and today he is still honoured throughout the country. His strategic grasp – clear-sighted, brave, yet pragmatic – deserves, however, a far wider appreciation. Perhaps his true genius in war, as Goethe famously said of writing, consisted in knowing when to stop.

GERD VON RUNDSTEDT

1875–1953

MICHAEL BURLEIGH

GERD VON RUNDSTEDT was 64 when the Second World War broke out, so with the exception of active service during 1914–18, most of his career was spent in peacetime. A quintessential Prussian army officer, his chief claim to fame was to have preserved and translated much of the ethos and spirit of the old imperial-Prussian army across the Weimar Republic into the Nazi era.

Yet this was illusory. He was a traditional Prussian military potentate, in his carmine striped trousers and much-decorated tunic, disdaining the new creed of leading from the front, and preferring to use large-scale 1:1,000,000 maps to get the grand strategic overview. But much of the genius even his enemies ascribed to him was not his: although he had the external symbols of military power, the rank and titles, in fact the real thing was invariably wielded by others. Behind the easy grand manner and the diplomatic expertise, it could be argued that Rundstedt was a successful sham. Liddell Hart was not alone in being taken in by Rundstedt's gentlemanly manner, imagining there was more than met the beholder's charmed eye. What Rundstedt actually presided over was the progressive nazification of the German armed forces, the final capitulation by the old elites that the conservative dissident Ewald von Kleist-Schmenzin had foreseen in 1934.

The long apprenticeship of a Prussian warrior

Rundstedt came from a distinguished Junker family with a history of military service to the Prussian crown. After graduating from the cadet academy at Gross Lichterfelde, in 1893 he became a lieutenant in the 83rd Royal Prussian Infantry Regiment. After a decade as an adjutant, he entered Berlin's prestigious War Academy in 1902. Only one in five of those who took the three-year course survived to embark on an eighteen-month probationary period on the General Staff. He succeeded in becoming a captain on the Troop General Staff. In that capacity he served as operations officer to the 22nd Reserve Infantry Division within the First Army, and participated in the invasion of Belgium and France in August 1914. The following year Rundstedt, now a major, was transferred to the more mobile Eastern Front, before being assigned to help rebuild the army of Austria-Hungary after its shattering defeat during the Russian Brusilov offensive. In 1917 he became chief of staff of LIII Corps, which advanced towards Petrograd so as to pressure the Bolsheviks into accepting the onerous Treaty of Brest-Litovsk. From March 1918 onwards he was chief of staff to XV Corps, which took part in the last major German offensive on the Western Front.

Given his connections and experience it was unsurprising that after the war Rundstedt should be chosen by General von Seeckt to join the Truppenamt, the Weimar Republic's disguised substitute for the General Staff, which the Versailles Treaty had abolished. He joined a distinguished group of newly minted lieutenant colonels, including Werner von Blomberg, Fedor von Bock, Kurt von Hammerstein-Equord and Wilhelm von Leeb, who dominated the Reichswehr under Weimar. During that period, Rundstedt advanced to the rank of major-general. By now in his fifties, he proved an adroit operator in the authoritarian reconstruction of the later Weimar Republic associated with the names of Hindenburg, Schleicher and Papen, emerging as the general commanding the Berlin-based First Army Group, the six divisions that defended Germany's eastern frontier.

Operating under the Nazis

The advent of Hitler's chancellorship in January 1933 resulted in the appointment as minister of defence of Werner von Blomberg, who in turn attempted to promote General Walter von Reichenau to the post of army commander-in-chief. Leeb and Rundstedt, who objected as much to Reichenau's relative youth as to his pronounced Nazi sympathies, stymied his appointment by threatening to resign. Rundstedt was only marginally less satisfied by the appointment as commander-in-chief of Werner von Fritsch, who had been a former subordinate and was also younger by five years.

Although Rundstedt often privately expressed his disdain for the Nazis, he and his colleagues acquiesced in the Night of the Long Knives in June 1934, in which Hitler ordered the murder of his rival, Ernst Röhm, and the other leaders of the Sturmabteilung (SA), which was threatening to become more than a paramilitary formation. Rundstedt and his colleagues also, apparently, acquiesced in the cold-blooded murder of General Schleicher and his wife, and Major-General Ferdinand Eduard von Bredow, who were all gunned down in the same SS mafia-style operation, which was logistically facilitated by the regular army. Rundstedt further demonstrated his loyalty by personally administering the oath that senior officers were obliged to swear to the Führer. In the same year as the Night of the Long

Knives, Blomberg, Fritsch and Rundstedt took prominent positions in the Nazi Party's annual Nuremberg rally. Rundstedt also undertook many of the social functions that fell upon General Fritsch, who as a bachelor workaholic had no time for such things as a courtesy appearance at the funeral of King George V.

Rundstedt managed to preserve the framework of the military old guard as Hitler expanded the German armed forces, even managing to tolerate such daring innovations as Guderian's infatuation with tanks. He saw himself as the embodiment of the army, saying 'we' think or want this or that in the first person plural. In early 1938 Rundstedt was afforded another glimpse into the mindset of his political masters. Blomberg was ruined after he had been encouraged to marry a woman who turned out to have a police record, while General Fritsch was simultaneously exposed as a closet homosexual. This last charge was trumped up and involved deliberately mistaking Fritsch for someone else. In a late-night interview with Hitler, Rundstedt demanded a court of inquiry to exonerate Fritsch, while refusing to countenance Reichenau as Fritsch's replacement. In the end, he agreed on General Walter von Brauchitsch. Fritsch was subsequently exonerated, insisting via Rundstedt on fighting a duel with Heinrich Himmler, whose subordinate Reinhard Heydrich had orchestrated the charges against him. The challenge was never delivered.

During the Munich crisis of late summer 1938 Rundstedt joined other senior generals in warning Hitler that the armed forces were not ready for a European war, although he coldly declined to join in their plans for a coup to overthrow him, describing the plot as 'base, bare-faced treachery'. After helping to occupy the Sudetenland in October 1938, Rundstedt retired with the honorary rank of colonel-in-chief of the 18th Infantry Regiment. But when Hitler set his sights on Poland in 1939, Rundstedt was recalled to undertake the military planning.

Case White: the invasion of Poland

In April 1939 Rundstedt, then aged 64 and officially retired, was given his own 'Working Staff Rundstedt', which operated from his home in Kassel. He was joined in this by Erich von Manstein and Gunther Blumentritt to prepare Case White for the invasion of Poland. By August Rundstedt's little group had mutated into a much larger operational headquarters for Army Group South, then massed on Poland's border. Rundstedt seems to have imagined that this was another example of Hitler's penchant for gambling for high stakes – another bluff or double-bluff like the one a year earlier when the German army had been poised to invade Czechoslovakia before the Munich agreement. But this time it was no bluff. Poland was rapidly enveloped as two armies, one moving southwards from East Prussia and the other northwards from Silesia, thrust towards Warsaw, the aim being to trap the main Polish force west of the Vistula. With a few brilliant modifications necessary to combat unexpected Polish resistance, Rundstedt and Bock's army groups invested Warsaw. After Soviet forces had invaded the country from the east on 17 September, Hitler decided to bring the capital quickly to its knees by launching a savage artillery and aerial bombardment. Warsaw surrendered on 28 September, the Polish government having already fled.

The attack in the west

After the spectacular success of the Polish campaign, Rundstedt spent a brief spell as military governor of Poland. But after the northwestern 'Warthegau' province and West Prussia were incorporated into Germany and the remnant turned into the 'General Government', Rundstedt was reassigned to the headquarters of Army Group A, one of the commands in the west. Like his colleagues, notably Brauchitsch and Halder, Rundstedt had little enthusiasm for the proposed western offensive against what they took to be formidable Anglo-French forces. He nevertheless silently acquiesced in Hitler's desire to have a war at all costs and his megalomaniac confusion of his own nihilistic destiny with that of the German people. Rundstedt was thus no more than a professional hireling whose job was to turn these fantasies into reality. In the existing version of the war-plan, Case Yellow, little more than a stalemate would ensue after the Germans had ploughed across Holland and Belgium to face British and French forces dug in along the line of the Aisne and Somme. This struck both Hitler and Manstein, who reappeared at Rundstedt's side, as both woefully unambitious and liable to result in high numbers of unnecessary casualties. Rundstedt backed Manstein's more ambitious scheme for a huge armoured thrust through the Ardennes, plans that gave substance to the Führer's own vision. However, Rundstedt also acted as a break on the more headstrong Manstein and Guderian by factoring in more armour and appointing an old-style cavalry commander, Ewald von Kleist, to command the armoured group.

As the invasion of France got under way in May 1940, Rundstedt was a constant cautious presence, insisting that the armour should not advance too rapidly, while always conscious of the need to secure the German flanks against the French commander Gamelin's possible counter-attacks. Hitler was sufficiently impressed to let him have his way. The Luftwaffe, rather than Rundstedt's armour, would take on the task of obliterating the British evacuating Dunkirk, while Rundstedt and Bock would concentrate on pushing ahead towards Paris. By 17 June Philippe Pétain was ready for an armistice. A couple of days later an exhilarated Hitler announced the promotion of twelve generals, including Rundstedt and his rival Reichenau, to the rank of field marshal. From their headquarters at Fontainebleau, the German General Staff turned its collective imagination to Operation Sealion, the invasion of Britain, it being envisaged that Rundstedt's Army Group A would seize a 100-mile bridgehead along the marshes and levels of Kent and Sussex. This was dropped in favour of seeing what Goering's Luftwaffe could achieve in the skies of Britain. In the event, the German Few fell like flies to the larger Few of the Royal Air Force in the Battle of Britain (July–September 1940).

Barbarossa and Army Group South

While other senior colleagues were relocated east, Rundstedt was made commander-in-chief in the west, in charge of all Wehrmacht forces in Belgium, France and Holland. He may have savoured the fact that Field Marshal Reichenau was now his subordinate. On 31 January 1941 Brauchitsch informed Bock, Leeb and Rundstedt that they were the designated group commanders for Operation Barbarossa, the greatest land invasion the world has ever seen. Rundstedt was to lead Army Group South, operating between the Pripet Marshes and the Crimea. The aim was to encircle and destroy Russian forces near the border, where neither new post-1939 defences had been built nor the old 1918 ones dismantled. Rundtstedt began marshalling his enormous forces on the Polish border. At a meeting of senior officers on 30 March 1941 Hitler made it clear that they would be waging a war to exterminate a political system, and that their 'quaint' notions of chivalry no longer applied. Rundstedt's reaction to this was unrecorded, as was his response to the matrix of 'criminal orders' that set the moral parameters for this campaign. Soviet political officers were to be summarily executed, while German troops were absolved of any responsibilities towards the Soviet civilian population. The army was to extend full cooperation to the SS *Einsatzgruppen*, which flitted about with the sole purpose of killing Jews. There is no evidence that Rundstedt demurred.

Barbarossa commenced in the early hours of Sunday 22 June, after Rundstedt and his two colleagues had signalled the codeword 'Dortmund' to the three army groups. Forty-three German and fourteen Romanian divisions jumped off as Army Group South; their opponents comprised eighty-nine Soviet divisions in two army groups. Since Stalin imagined that Hitler would concentrate on the economic resources of the Ukraine, rather than the political target of Moscow, he had focused his forces in the south and sent General Georgi Zhukov, one of his top commanders, to oversee them. A direct result of this was that Rundstedt had to scale down plans for a vast encirclement of Soviet forces in favour of the rapid seizure of Kiev and the trapping of a relatively small Soviet force after the main one had escaped. The surrender of 100,000 Russian troops was small beer in comparison with what took place elsewhere during Operation Barbarossa.

After Hitler had decided to concentrate on economic goals in the north and south, Rundstedt's southern army group pressed southwards towards the Crimea and into the area between Kharkov in the north and Rostov-on-Don in the south, the gateway to the oil of the Caucasus. Hitler's SS bodyguard, the SS-Leibstandarte, took Rostov on 21 November, although they were forced out by a Russian counter-attack and the Germans had to fight their way back in once more. When Rundstedt expressly contemplated withdrawal, Hitler replaced him as army group commander with the more biddable Reichenau. Ill after a heart attack, Rundstedt experienced that rare thing, an apology from Hitler, after Sepp Dietrich, commander of the SS-Leibstandarte, admitted to Hitler that the only alternative to retreat from Rostov would have been the annihilation of its four divisions. Rundstedt declined to spend the large cash sum that accompanied the apology. In January 1942 he had the satisfaction of representing Hitler at Reichenau's funeral after his rival had suffered a fatal stroke.

Defending Fortress Europe

In the summer of 1942 Rundstedt was appointed to command Army Group D, based in the occupied zone of northern France, and also given overall command of the whole of the western theatre. Rundstedt managed to sustain good relations with Pétain – he spoke good French – even as he carried out Operation Anton (November 1942) in which the Germans occupied the whole of Vichy France as a precaution against an Allied landing on the Mediterranean coast.

He also oversaw the construction of the vast Atlantic Wall, designed to repel invaders from the west, while at the same time suffering from a constant depletion of his own forces to sustain Germany's flagging efforts on the Eastern Front, a process he warned would result in a worrying inferiority vis-à-vis the Allied forces massing in England for a cross-Channel invasion. In return for men sent east to fight the Soviets he received low-grade 'eastern troops' – Ukrainian and Russian conscripts and collaborators who made up one in six of 'German' forces in France. Many German units also consisted of men aged over 30 or with similar categories of wound or ailment, so that entire battalions had men with stomach problems. The size of a German division had also shrunk from an average of 17,000–18,000 men earlier in the war to a notional 13,000. To correct all this, in November 1943 Hitler dispatched the dynamic Erwin Rommel, at 51 the youngest German field marshal and eighteen years Rundstedt's junior – Rundstedt referred to him as 'Marschall Bubi', roughly meaning 'Marshal Laddie'. The two men fundamentally disagreed as to whether it would be better to mass German armour near the beaches, so as to wipe out the Allies while their boots were still wet, or to keep armour in reserve to be deployed in a major engagement should the Allies break out from their beachheads. This last strategy left the tank forces vulnerable to Allied air attack should they seek to move any distance. Rundstedt was then effectively made Hitler's cipher when the dictator assumed overall command of the army in the west.

When the invasion came in June 1944, the German defences were overstretched, having been deceived by the Allies into thinking that the landing would take place further north. After losing the argument with Hitler about how to regroup to combat the invasion, Rundstedt received a letter regretfully accepting his (spurious) resignation on health grounds. The addition of Oak Leaf Clusters to his Knight's Cross was supposed to soothe the sting of this rude dismissal.

Within weeks, in the wake of the failed July bomb plot, Rundstedt was back in harness, presiding over the court of honour that discharged fifty-five senior officers allegedly involved in the conspiracy so that they would be liable to a degrading execution after 'trial' before Roland

Friesler's People's Court. Many of his colleagues never forgave Rundstedt for playing this part in Hitler's gruesome revenge, and there was more than a whiff of hypocrisy as Rundstedt delivered the eulogy after the enforced suicide of his colleague Erwin Rommel, who had also been implicated in the plot. In September, in return for these political services, Rundstedt was reinstated as commander-in-chief in the west, not at Fontainebleau, but at Arensberg outside Koblenz, a dismal reminder of the pass to which Hitler had brought Germany. Although exhaustion and overextended supply lines accounted for the Allies' sudden halt in October 1944, even the Allies' own popular press attributed this to Rundstedt's skills as a generalissimo, greatly exaggerating the one undoubted German victory achieved by Model at Arnhem, when he temporarily routed a combined airborne operation. The Allies also attributed to Rundstedt the initially devastating Ardennes offensive (16 December 1944 – 25 January 1945), although he had little or nothing to do with its conception. By March 1945, with US forces pouring across the Rhine at Remagen, Hitler decided that Rundstedt was too old for the job and replaced him with Field Marshal Albert Kesselring.

Rundstedt was captured by US troops on 1 May 1945 while recuperating in a hospital based at the SS training school of Bad Tolz in Bavaria. He claimed that only lack of fuel and airpower had stopped him repelling Operation

Overlord, the Allied invasion of northwest Europe. Rundstedt was imprisoned in the 'Ashcan' holding centre for key captives near Spa in Belgium before being transferred, with his English-speaking son Gerd for company, to Camp 11 at Bridgend, near Swansea, in Wales. He was able to convince the Allies that, despite his rank, he was not a key player in the decision to go to war or in deciding how that war was conducted. Although he avoided the tribunal for major war criminals, he was kept in captivity for four years as the Allies pondered lesser charges. He was freed on health grounds in 1949 and settled in a flat above a shoe shop in Celle. He died on 24 February 1953; the officiating cleric spoke of 'the burial of the last great Prussian'.

ERICH VON MANSTEIN

1887–1973

MICHAEL BURLEIGH

ERICH VON MANSTEIN was one of Nazi Germany's leading generals. He was the architect of Hitler's lightning victory in the west in the summer of 1940, and responsible for a number of brilliant holding operations on the Eastern Front, operations that have received less attention than the more dashing offensives of other generals. Although Manstein did more than most, through his memoirs, to shape the image of the decent Wehrmacht's studious non-involvement in Nazi war crimes and crimes against humanity, nowadays he is thought to have been heavily implicated in both. The new Manstein is a grim figure, far removed from the much-decorated noble knight of his own imagining.

Manstein was born to command. Fritz Erich von Lewinski was the tenth child of a Prussian artillery general who died during a training accident. Thereafter he was brought up by an uncle, General Georg von Manstein, and his wife Hélène, who were childless. Erich von Manstein, as the boy became, had soldiering in his blood; some sixteen of his ancestors had been generals. One of his uncles was Field Marshal (and future president) Erich von Hindenburg. Manstein progressed swiftly from the prestigious Prussian Cadet Academy to the Imperial War Academy, before serving as a guards officer during the First World War. After being badly wounded, he spent the rest of the war as a General Staff officer on both the Eastern and Western Fronts. By this time he had fully imbibed the ethos and values of the Prusso-German military caste.

Manstein's political views

Generals are never impervious to the politics and prejudices of their age and class, least of all in unsteady democracies or dictatorships. Manstein's military career continued unabated as a Reichswehr officer under the Weimar Republic and then, after January 1933, as an officer in Hitler's Wehrmacht. As a former royal page, Manstein regarded the abdication in 1918 of the Kaiser as the 'collapse of his world' and the Versailles Treaty as an 'act of dishonour'. He detested the party politics of Weimar and welcomed the authoritarian turn that Hitler represented. Although he disliked the purging of officers of Jewish ancestry, since for him caste was everything, Manstein welcomed the broader anti-Semitic thrust of the Law for the Protection of the Professional Civil Service that these measures came under.

By 1936 Manstein had become quartermaster general, traditionally the stepping-stone to becoming Chief of the General Staff, in this case in order to follow Ludwig Beck, and such distant luminaries as Moltke and Schlieffen. Perhaps because of his closeness to disgraced General Werner von Fritsch, Manstein was abruptly posted to command the 18th Infantry Division based in Silesia, although he was careful to maintain his General Staff connections. There he delivered a speech celebrating Hitler's fiftieth birthday, which expressed approval of all the measures the Führer had taken since 1933 and the aggressive revision of Versailles that Hitler envisaged. Manstein fully subscribed to the view that Germany had been deliberately 'encircled' by the Allies, and eagerly supported Hitler's intention to break out of this encirclement with a more robust foreign and security policy.

Manstein played a major role in the illegal German annexation of Austria and the incorporation into the Wehrmacht of its armed forces, and then in the occupation of the Sudetenland in the wake of the Munich agreement of September 1938. During the invasion of Poland, Manstein served as chief of staff under Field

Marshal Gerd von Rundstedt, the commander of Army Group South.

The planner

Manstein played a major part in overhauling the outmoded plans for the invasion of France in the summer of 1940 after he was posted to Koblenz in late October 1939. The existing plan, Case Yellow, was a degraded version of the elaborate First World War Schlieffen Plan. Instead of seeking the enemy's encirclement on a vast scale, Case Yellow projected losing up to half a million men in a broad and crude frontal attack designed to hurl the Allies across the Somme. After these losses it would take a further two years for Germany to recover sufficiently to launch an onslaught against France. The strategic pessimism reflected in this plan may have reflected the desire of key generals to postpone the western invasion at a time when they felt Hitler was leading Germany to disaster. By contrast, Rundstedt's thought was more in line with Hitler's own love of the decisive gesture. To that end he commissioned Manstein to introduce more movement and surprise into projected operations. In this, he succeeded magnificently.

The Fall of France

Backed by Rundstedt, but opposed by Army Chief of Staff Halder, Manstein proposed a plan that Churchill subsequently dubbed the 'Sickle Cut'. Its crucial innovation, which may have been due to a chance intervention by the tank expert Heinz Guderian (see pp. 193–207), was to push armoured formations through the inhospitable terrain of the Ardennes and Eifel, so as to encircle and destroy the main Allied forces on the Channel coast. There were three major thrusts, launched during a spell of such fine weather that Hitler rewarded his chief meteorologist with a gold watch.

In the north Fedor von Bock's Army Group B 'waved its matador's cloak' so as to lure the British Expeditionary Force and the French army in Flanders northeastwards into Belgium, thereby leaving them vulnerable in the rear. To the south, General Leeb's Army Group C would pin down French forces defending the Maginot Line. Lastly, in the centre the forty-four divisions of Rundstedt's Army Group A would traverse the Ardennes, which the French had lightly garrisoned with poor-quality troops, with a view to emerging behind the Allies in the region of Boulogne. Among the factors leading to Allied defeat were the presence of such generals as Guderian and Rommel, and the Luftwaffe's successful coordination of bombers and Stukas with the advancing armoured formations. As a result of these operations, the British were forced to evacuate through Dunkirk, while German forces ploughed forwards and took an undefended Paris on 14 June. Contrary to

Halder's gloomy prognostications, it had taken Germany only six weeks and the loss of nearly 30,000 men to take France. Notwithstanding the monumental traffic jams and the reliance on infantry and horses rather than the armoured blitzkrieg of legend, Manstein had been largely responsible for planning this brilliant operation.

Fearing that Manstein was part of a ploy by Rundstedt to achieve operational autonomy, Halder engineered Manstein's posting as commander of the 38th Infantry Corps, even as Hitler approved Manstein's modified version of the plans for the invasion. Halder was given the task of working Manstein's ideas into a fresh master-plan. Three million German troops were subsequently deployed for the invasion, a quarter of them veterans of the First World War aged over 40. Shortly after this victory, Manstein was made an infantry general and awarded the Knight's Cross to add to his Iron Cross. Even his opponents began to refer to him as a 'military genius', partly so as to cover up their own egregious errors.

Thereafter, Manstein was involved in the planning for Operation Sealion, the proposed German invasion of Great Britain, and the occupation of northern France.

Manstein and Operation Barbarossa

Plans to invade Britain petered out as Hitler convinced himself that the solution to Britain lay in the plains of

Russia. In February 1941 Manstein was transferred to command the 56th Motorized Armoured Corps, operating on the Leningrad front. In that capacity his tanks ventured as far as Lake Ilmen, leaving their supply train some 90 miles behind. This led him to be considered a master of wars of movement. What received less attention was that his Operation Northern Light was unsuccessful in taking and destroying Leningrad. Instead, he recommended investing the city and starving its defenders and inhabitants to death. The total absence of moral scruple is noteworthy in a man whose self-image was that of an apolitical soldier – unlike the dastardly ideologues of the SS.

Although he had no direct role in establishing the wider ideological framework for the conduct of the war on the Eastern Front, Manstein was responsible for implementing the various orders that contributed to the barbarous behaviour of German operations in this vast theatre. Following the death of General von Schobert in a plane crash in September 1941, Manstein was appointed commander-in-chief of the Eleventh Army, which was the most southerly part of Army Group South. Manstein's role was to conquer the Crimean peninsula. In this capacity, he relayed the 'Reichenau Order', which enjoined the Wehrmacht to collaborate in crimes against the Soviet civilian population. This especially applied to the Jews: 'The Jewish-Bolshevik system must be annihilated once and for all. It must never be allowed to interfere in our European living space.' The practical result of this was that Manstein's Eleventh Army fully cooperated with the mass murderers under SS Gruppenführer Otto Ohlendorf of *Einsatzgruppe D*, who slaughtered approximately 90,000 Jews, Gypsies and psychiatric inmates in the course of their operations in the Crimea, relying on the Eleventh Army for both intelligence and logistics. Regular soldiers set up perimeter cordons and were occasionally also involved in the shooting of victims. Ohlendorf would subsequently testify at Nuremberg that relations between the *Einsatzgruppen* and the Wehrmacht were excellent. One document shows how Manstein personally ordered the distribution to his men of wrist-watches looted from 10,000 Jews murdered by Ohlendorf's men in Simferopol. He also vainly tried to reconcile insistence that his troops live from local resources with the political imperative of rallying all non-Bolshevik forces, arguably the single greatest blunder of the entire campaign on the Eastern Front.

From Sebastopol to Stalingrad

Ably assisted by the Romanians, in 1942 Manstein launched a second assault on the harbour fortress at Sebastopol, after he had connived at the downfall of General Count Sponeck by refusing to permit him to withdraw to evade a counter-attacking Soviet pincer movement. In this respect, Manstein behaved exactly like Hitler, who wanted Sponeck shot. In the event, he was tried by court martial and sentenced to six years in jail. Using heavy 60 cm and 80 cm mortars, and heavy bombing raids by the Luftwaffe, in July 1942 the Eleventh Army succeeded in expelling the Soviets from Sebastopol. Manstein was promoted to field marshal. In August he was transferred north to command the ongoing siege of Leningrad: his role, to repulse Soviet attempts to relieve the besieged city.

Against the background of the developing crisis at Stalingrad, in November 1942 Manstein was given command of a new Army Group Don, which included General Paulus's beleaguered Sixth Army, surrounded at Stalingrad by the Soviets and with its supply lines too extended. A relief operation (Winter Storm) was designed to come to the aid of the hapless Paulus, who had never commanded such large formations, let alone under such atrocious conditions. General Hoth would drive his armoured column to within 20 miles of the city, linking up with a force that had managed to break out. In the event, Hitler refused to countenance the break-out until it was too late for it to succeed, and Hoth remained stuck some 30 miles from Stalingrad. The fate of Paulus's Sixth Army was sealed, and the survivors were marched off to camps in Siberia.

Manstein had long sought to introduce more movement into the campaigns on the Eastern Front, while seeking to consolidate the army's command structure under a Chief of General Staff and a commander-in-chief for the Eastern Front, a role he envisaged for himself. This was intended to reduce the role of Hitler in day-to-day military decision-making and led to growing tensions between the two men. Manstein was also furious that while half the German army was scattered across western Europe waiting for invasions that failed to come, the other half was being hard pressed by the Soviets. Frustrated with Hitler's continuing interference, Manstein trained his pet dachshund Knirps to do the 'Hitler Greeting' with his right paw whenever the dog heard 'Heil Hitler'.

Managing insoluble positions

Great commanders are not always the hard-chargers who dash in to destroy their enemy or occupy his territory. During early 1943 Manstein played a distinguished part in counter-offensives to retake Kharkov and Belgorod, for which he was awarded oak leaves to his Knight's Cross. Although his relations with Hitler were increasingly fraught, especially after Hitler promised and then failed to deliver reinforcements, Manstein took no part in the military conspiracy to assassinate the Führer in the summer of 1944, finding it impossible to reconcile this deed with his soldierly conception of honour. When Claus-Schenk von Stauffenberg elliptically broached his discontent with the Führer,

Manstein offered to transfer him to the front in order 'to clear his head'. He also rebuffed an offer to become the conspirators' future Chief of Army General Staff, an offer made by Rudolf-Christian Freiherr von Gersdorff of Army Group Centre, with the riposte: 'Prussian field marshals do not mutiny!'

Strategic withdrawal after Kursk

In the summer of 1943, Manstein led Army Group South in an attempt to pinch off the heavily defended Kursk salient, an operation so vast in scale that Hitler said it made his stomach turn even thinking about it. Four million troops were involved, along with 13,000 armoured vehicles and 12,000 planes. Manstein clashed with Hitler over the timing of Operation Citadel, as the German attack was codenamed. Hitler wanted it delayed to incorporate new weapons systems, but this merely enabled the Red Army to fortify the salient in prodigious depth. After the Kursk offensive was abandoned, and ignoring Hitler's orders to stay put, in September 1943 Manstein effected a brilliant strategic withdrawal of three of his armies across a 450-mile front. This involved using six bridges to cross the Dnepr, the largest river in Europe. Having executed this manoeuvre, he then counter-attacked around Zhitomir to stabilize the shortened front.

Withdrawal and stabilization were not what Hitler

sought. On 30 March 1944 Hitler relieved Manstein of his command of Army Group South, fobbing him off by promising him command of a future western offensive should it come. A more concrete consolation was the gift of a large landed estate, although that could not compensate him for the loss of his eldest son on the Eastern Front. Manstein was relieved of his post and never received another command. He was arrested and interned by British forces in August 1945. Allied courts and the general himself raced to shape the immediate past with a view to how posterity would remember it.

On trial

Manstein was initially summoned as a witness for the Allied prosecution at Nuremberg. He was one of the senior German generals who were required to produce a memorandum clarifying the role of the military under the Nazi dictatorship. The memorandum managed to dispense with such vulgar notions as guilt or responsibility for crimes that by then were well known. Old mentalities were unshakeable. Writing to his wife, Manstein highlighted his own part in ensuring that the Wehrmacht was not condemned as a criminal organization like the SS. This left unexplored the question of who had issued the complex of so-called 'criminal orders' that set the moral parameters on the Eastern Front by licensing mass murder. Manstein's memorandum set in motion the legend that the Wehrmacht had fought a clean and honourable war, despite the worst efforts of the SS to besmirch the image of the German fighting man. This myth largely took hold, despite the fact that between 24 August and 19 October 1949 Manstein himself faced seventeen counts of war crimes before a British military tribunal. Manstein's defence was helped by Soviet efforts to extradite him to face similar charges in Russia, and by Basil Liddell Hart's book *On the Other Side of the Hill*, which included a number of German generals' accounts of the war.

Manstein skilfully played to Western apprehensions in the early Cold War by talking up how the 'Asiatic' character of the fighting affected both sides. He claimed to have identified this as early as 1931 when he observed Red Army manoeuvres under the terms of military cooperation between the USSR and Weimar Republic. Manstein had much public sympathy in Great Britain, Churchill himself feeling sufficiently moved by the general's plight personally to contribute money to his defence fund. Manstein was sentenced to eighteen years in prison for ordering the shooting of Soviet political commissars and prisoners of war, and the abduction and killing of civilians. There was no mention of his role in facilitating *Einsatzgruppen* operations against Jews, Gypsies and the mentally ill. The sentence was reduced to twelve years on appeal; he was released on health grounds on 7 May 1953.

Post-war role

During the 1950s Manstein played a prominent role in advising on the formation of the West German Bundeswehr, while issuing a stream of letters and publications designed to exonerate the wartime Wehrmacht from charges of organized criminality. He died in 1973, having successfully shaped how future generations would think about the war, a feat he achieved through calculated amnesia regarding some of its most horrendous features and by narrowing notions of war down to the conduct of individual commanders and operational issues.

Manstein's role as a planner of the brilliantly successful invasion of France in the summer of 1940 continues to eclipse his indifferent conduct during the defeat at Stalingrad and the defeat of the German offensive at Kursk. Not only did he not stand up to Hitler, but he was clearly involved in facilitating Nazi war crimes, as well as barbarous behaviour towards Soviet civilians. An earlier enthusiasm for the field marshal on the part of his former foes seems remarkably unfortunate. Basil Liddell Hart thought that 'The ablest of all the German generals was probably Field Marshal Erich von Manstein. That was the verdict of most of those with whom I discussed the war, from Rundstedt downwards.' Rundstedt was probably being characteristically modest.

HEINZ GUDERIAN

1888–1954

MICHAEL BURLEIGH

HEINZ GUDERIAN is chiefly remembered as one of the great practitioners, and theorists, of armoured warfare, in which tanks, motorized artillery and infantry with close supporting aircraft deliver swift crushing blows to enemy forces. As he once remarked with characteristic bluntness: 'You hit somebody with your fist, and not with your fingers spread!' Guderian first put this doctrine into practice in a remarkably executed motorized dash from Sedan to Abbeville and Calais during the invasion of France in the summer of 1940, an exercise he repeated on a vaster scale during the opening thrusts of Operation Barbarossa into the Soviet Union a year later. Less often remarked is that Guderian was a leading advocate of anti-tank warfare, attaching such dedicated units to his own formidable panzer divisions. He was also one of the few commanders who stood up to Hitler, and almost alone in being dismissed and recalled by the Führer.

Guderian admired the British military historian Basil Liddell Hart (see pp. 133–47), and the latter admired him back. But against Liddell Hart's laudatory view of 'Hurricane Heinz' or 'Speedy Heinz' must be mentioned Guderian's political involvements, the secret gifts and payments he received from Hitler, his alleged ignorance of the 'criminal orders' issued before Barbarossa, his involvements in an arms industry reliant on foreign forced labour, and a number of his own strategic decisions that resulted in disaster. These things have so far not overshadowed his image in the popular imagination as a dashing panzer general.

Heinz Wilhelm Guderian was born in 1888 at Kulm in West Prussia (now Chełmno in Poland), the son of a lieutenant general. He served in the cadets at Karlsruhe before joining Berlin's prestigious War Academy, and during the First World War he served as a signals officer on the Western Front, having taken specialized courses in radio communications. His belief in wars of rapid movement was largely an instinctive response to the paralysed war of attrition he witnessed among the trenches of Flanders. He experienced Germany's collapse in northern Italy, and saw demoralized and bolshevized soldiers in Berlin and Munich first hand. In January 1919 he was posted to a newly minted 'Eastern Frontier Force', intended to stop Russian and Polish incursions into his Prussian homeland. He was finally posted as a liaison officer between the paramilitary Freikorp's Iron Division and the General Staff.

Views on armoured warfare

After the abandonment of the Baltic States to a precarious independence, Guderian entered the Reichswehr, becoming part of its covert General Staff, an institution proscribed by the Treaty of Versailles. Successive Reichswehr leaders sought to circumvent this and other restrictions on Germany's armed forces by modernizing them instead. In 1922 Guderian was appointed to investigate the possibilities of motorized warfare, to which end he studied the available technical literature, and (being able to read English and French fluently) monitored British experiments in using massed tank formations controlled by radio telephony – the British were pioneers of linking radios through a common frequency into what was called a 'net' work. By 1933 Guderian was a full colonel, convinced that in future conflicts rapid thrusts by independent armoured formations, commanded aggressively by leaders linked to tanks by radio, would be able to win the day by attacking the flanks and rear of the enemy or by chasing them when they fled. The fate of the formation's own flanks was an irrelevance given the mobility of these formations. His superiors had other ideas. When told that Guderian wished to lead 'from the front by wireless', General Ludwig Beck replied, 'Nonsense! A divisional commander sits back with maps and a telephone. Anything else is utopian.'

Legend has it that during an exercise involving Mark 1 tanks at Kummersdorf Guderian so impressed Hitler that he exclaimed 'That's what I need! That's what I want to have!' In fact, it is doubtful whether Hitler would have seen much potential in the glorified tractors or lorries with fake wooden turrets he saw that day. When it came to rebuilding the armed forces, Hitler favoured the Luftwaffe, and continued to think in conventional terms of artillery, infantry and cavalry, with armour only there in support.

Through sheer persistence, Guderian was able to force through building programmes that delivered light and medium tanks in sufficient numbers to make a plausible case for armoured warfare. He conducted the first exercise involving controlling tanks by radio telephony in 1935. In October of that year Hitler commissioned him to establish three tank divisions. A year later Guderian was promoted to major-general, and in 1937 published *Achtung – Panzer! The Development of Armoured Forces, Their Tactics and Operational Potential,* a book that propagated his theories of armoured warfare. Tanks, Guderian wrote, would be the spearheads of armoured battle groups. These would take pivotal positions, using anti-tank guns to ward off counter-attacking armour, before moving on for another thrust. The following year he was promoted to chief of mobile troops, commanding all tanks and motorized forces in the Wehrmacht, a mixed blessing since it was part of a ploy by the more senior 'Gunners', Commander-in-Chief of the Army Walther von Brauchitsch and General Ludwig Beck, to sideline him.

Field commander in Poland and France

In September 1939 Guderian commanded the 19th Panzer Corps in the invasion of Poland, his field of operations including his birthplace at Kulm. Fighting in difficult country, his tanks rapidly took their objectives, and Guderian was politically astute enough to advertise these triumphs when Hitler arrived on a tour of inspection. In eight days Guderian had advanced 100 miles to Brest-Litovsk.

Already distinguished by his role in Poland, Guderian played a crucial part in the reformulation of Case Yellow, the plan for the invasion of France, since he confirmed to his relative Wilhelm Keitel, head of the Wehrmacht High Command, that it was possible to take tanks through the intractable terrain of the Ardennes, where roads were few and of poor quality. At a final planning session, Hitler asked Guderian what he intended to do after he had crossed the Meuse. Guderian replied: 'I intend to advance westwards ... In my opinion the correct course is to drive past Amiens to the English Channel.' When the invasion came in May 1940, Guderian commanded a lead armoured formation in the thrust performed by Army Group B into France. There were constant rows as his more cautious superiors kept halting his forward advance in fear that such extended lines might be subject to attack on the flanks. One row led to his resigning on the spot, only to resume his command a few hours later when a compromise was reached, by which his tanks were allowed to carry out probing reconnaissance. Similar equivocations at the highest level prevented Guderian from delivering a killer blow to the Allied forces that had fallen back at Dunkirk. He blamed his senior rival, the mediocre Günther von Kluge, rather than Hitler for this fateful decision to spare the British from annihilation.

By way of reward, Hitler created a special Panzer Group Guderian, enabling the general to have giant 'Gs' painted on the vehicles of this two-corps armoured formation. He was promoted to colonel general, becoming one of the licensed idols in Goebbels's propaganda campaigns. A tough-looking man in a leather coat, Guderian was a master of pithy sayings, like 'Kick them, don't spatter them' or, as his tanks surged forward, 'Anyone for a ticket to the final station?' It was at this crucial juncture that Guderian claimed to have favoured a German Mediterranean strategy rather than an attack on Russia. According to this view,

Germany should have concentrated on driving the British from Gibraltar, Malta, North Africa and Suez, thereby perhaps forcing Churchill's resignation and the appointment of a government more amenable to a peace deal. After picking off the British, the Germans could then have turned on Russia while the USA stood by. But Hitler was intent on opening a war on two fronts, and, despite his reservations, Guderian ended up playing a key role in the overwhelmingly successful early stages of Operation Barbarossa in the summer of 1941.

Operation Barbarossa

Ironically, the master of fast-paced mechanized warfare was initially deeply sceptical about the advisability of Germany fighting a two-front war by invading the USSR. Guderian later recalled that 'When they spread a map of Russia before me I could scarcely believe my eyes.' He thought his superiors were mad in thinking they could defeat Russia in a couple of months, especially by relocating captured French tanks to conditions for which they were unsuitable. He went so far as to express his reservations in writing to the chief of staff, Brauchitsch, who ignored him. When he realized that Hitler was hell-bent on this strategy, Guderian immediately began an intensified training programme for his own armoured divisions, while pressing the armaments minister, Fritz Todt, for enhanced production of mechanized vehicles. Guderian demanded monthly tank production be raised from 125 to 800, then to 1,000 vehicles. This was impossible to fulfil in the light of Germany's other commitments.

When Operation Barbarossa began, on 22 June 1941, Guderian commanded his own independent panzer group, consisting of three panzer corps, within Field Marshal Fedor von Bock's vast Army Group Centre. Guderian's formation had some 850 tanks when it set off from Brest-Litovsk on its fifteen-day journey to the Dnepr in the heat and dust of Russia's endless plains. Beyond lay further rivers to cross, including the Desna and the Oka. The strategy of leading from the front meant that Guderian was almost captured by the Russians on the third day of the attack. Throughout the advance, Guderian and his superior Kluge constantly bickered, with Kluge flying in to Guderian's HQ to insist that he halt on the Dnepr so that the infantry could catch up. This time Guderian got his way, and his forces crossed the river, destroying Soviet forces around Mogilev and taking Smolensk on 15 July. (It was there that they seized the entire archives of the local Communist Party, which when eventually captured by the Americans, would shed unique light on the inner workings of the Soviet Union.)

Guderian forged ahead towards Gomel, encircling and destroying some ten Russian divisions along the way. However, by mid August, although millions of prisoners had been taken, there was no sign of an end to the millions of replacements that the Soviets were putting into the field. At this point Hitler made the fateful decision to abandon the political objective of Moscow in favour of concentrating on Leningrad, the Crimea and the Ukraine.

After mid August, when Hitler abandoned his plan to advance on Moscow, German momentum in the centre was halted in favour of a defensive posture, while all efforts were concentrated on the wings. Guderian endeavoured to remonstrate with Hitler to stick with the original objective, but found himself alone in a room with more senior generals nodding in agreement with their Führer.

In September Guderian took his tank forces southwards to encircle Russian forces defending Kiev. They captured 660,000 men. The ease with which that was accomplished persuaded Hitler to resume the thrust to Moscow, fatally underestimating the reserves he would need to bring up to accomplish this. Despite having only a quarter of the tanks that he had started off with in June, Guderian was a firm supporter of this renewed thrust. In October Hitler duly ordered the resumption of the attack on Moscow, at a time when the weather was already inauspicious and Stalin was bringing fresh troops from Siberia more than able to cope with the freezing temperatures descending on western Russia. For a time the Germans benefited from the element of surprise; when Guderian's tanks attacked Orel, the trams were still crawling along its streets.

But the seemingly unstoppable advance encountered stiff resistance at Mtsensk. There Guderian's formations were hard pressed by new T-34 Russian tanks, which outclassed his panzers on poor ground. He immediately ordered studies of captured T-34s and demanded anti-tank guns capable of destroying them. At Tula, Guderian again encountered fierce resistance. In letters to his wife he wrote of the atrocious weather, the inadequate denim clothing worn by his men, and the paralysing lack of fuel, as his logistics lines stretched further and further.

By the opening days of December Guderian was forced to withdraw as the 1st Guards Cavalry Corps counter-attacked. Flying back to Hitler's headquarters at Rastenburg in East Prussia, Guderian pleaded for a tactical withdrawal. 'Positional warfare in this unsuitable terrain will lead to battles as material as in the First World War,' he explained. 'We shall lose the flower of our officer and NCO corps. We shall suffer huge losses without gaining any advantage. And these losses will be irreplaceable.' Having earlier praised Guderian's hands-on approach to leading from the front, Hitler now said his angle of vision was too restricted to the front. German troops came within 18 miles of central Moscow before the attack petered out through lack of reserves and matériel. At minus 34 degrees centigrade it was so cold it was impossible to dig in, while motor oil congealed.

Germany had suffered its worse setback since 1918, and a defeat of such magnitude that it permanently crippled its prospects of victory in Russia. Guderian was partly responsible for that fateful decision. He imagined that the Soviets were at the end of their capacities and were scraping the barrel to put new forces together. In fact they had an almost infinite supply of fresh manpower, and had relocated their industries eastwards beyond the Urals to churn out sturdy no-nonsense weaponry to fight the more fancily equipped Germans – who would soon suffer critical shortages of spare parts as well as of men and fuel.

Dismissal and return

On 24 December 1941 Guderian's arch foe Kluge accused him of an unauthorized withdrawal that had inadvertently left a gap of 25 miles between German positions. The following day – Christmas Day – Hitler dismissed Guderian from his command. By January 1942 the Wehrmacht had suffered nearly a million losses – wounded, captured, missing or dead – or about 29 per cent of the 3.2 million men involved in the opening phase of Barbarossa. By then the Wehrmacht had been pushed back nearly 200 miles west of Moscow, and the Soviet leadership had quietly returned from whence they had fled.

After recovering from a heart attack, Guderian spent the next fourteen months inactive. In October 1942 Hitler gave Guderian a sum of money sufficient for him to purchase Dneipenhof, an estate in the so-called Warthegau area of occupied Poland from which the original Polish owners were evicted. This was in addition to the 2,000 Reichsmarks a month Guderian received from Hitler to top up his officer's pay, not to mention the family estate at Gross-Klonia, also in the Warthegau, which the Guderians had retrieved in 1939.

Guderian was recalled on 1 March 1943 as inspector general of armoured troops, reporting directly to Hitler. Within a month of his appointment he had raised the monthly production target from 600 to 1,955 tanks. Together with Speer he raced against the clock to take new tanks like the Tiger through their teething troubles or to 'up-gun' existing vehicles. He spent much time touring arms factories, making informed suggestions as to how to improve design from the point of view of the tank crews – he knew how to drive a tank and to fire its guns. During none of these tours of arms factories did Guderian appear to notice their mounting dependence on foreign forced labour. Simultaneously he tried to dissuade Hitler from such follies as prematurely attacking entrenched Soviet defences at Kursk.

Chief of the General Staff

Having rebuffed one set of military conspirators in 1942–3, Guderian declined to take part in the July 1944 attempt to assassinate Hitler. He spent 20 July, the day Stauffenberg's bomb failed to kill the Führer, shooting roebuck on his estate, ensuring that he remained incommunicado. In the gruesome aftermath, he was appointed to the Court of Honour that cashiered the surviving plotters prior to their trial before a People's Court. In return for his loyalty, Hitler appointed Guderian Chief of the General Staff, although the status of the position had been much degraded. In his new role Guderian gave a radio broadcast in which he called for a National Socialist officer corps led by General Staff officers who should 'exhibit the thoughts of the Führer'.

Despite this public loyalty, there were epic clashes between Guderian and Hitler, under the silent gaze of the ambient sycophants at Hitler's headquarters. Again and again Guderian told Hitler to evacuate troops trapped in the Courland pocket, and to deploy them elsewhere. Hitler refused, claiming they were holding up larger Russian forces. Guderian also realized that the Ardennes offensive – an attempted re-run of his own earlier triumphs – was misconceived, and tried to have it broken off when it failed, so as to reinforce the crumbling Eastern Front. There were endless rows about the appointment, performance, dismissal or promotion of various generals – questions of personality that frittered away valuable time amidst gossip and innuendo.

Things came to a head on 28 March 1945, when Hitler insisted on dismissing General Theodor Busse for failing to defend Kustrin. He inveighed against the army, the General Staff, the officer corps and the generals. Guderian responded by criticizing Hitler's military leadership and his heartless abandonment of the inhabitants of eastern Germany to the invading Red Army. He insisted that a senior army figure be appointed to Himmler's incompetent SS staff. Hitler was so incandescent with rage that onlookers thought he might physically hit the general. After being restrained, Hitler said 'Colonel General Guderian, your physical health requires that you immediately take six weeks' leave.' By the time that period had elapsed, Hitler had committed suicide. At the end of the war Guderian was taken into captivity by the Americans. The Poles tried to extradite him, but he never faced war crimes trials. He was released from US custody in 1948. Guderian died on 14 May 1954 at the age of 65.

Assessment

Guderian was one of the most important advocates of armoured mechanized warfare, and a distinguished commander of such forces in the field. Whereas Hitler initially regarded such formations as part of a gigantic bluff to convey the impression of huge armed might, Guderian wanted to build them up into the real thing – a formidable fighting force capable of smashing through the enemy's weak points. With dash and flair bordering on recklessness, he was one of the commanders responsible for Germany's victory in the west in the summer of 1940. A year later he essayed similar rapid advances over the much greater expanses of Russia, despite fully realizing the likelihood that these would simply peter out as the bitter Russian winter set in and the enemy's enormous potential resources came into play.

Guderian was certainly not afraid to speak his mind to Hitler, let alone to his own commanders, but this sits curiously with the fact that he was happy to accept covert bribes and pay-offs, which more honourable figures like Rommel declined. There is also the disturbing detachment in his memoirs regarding unspeakable atrocities committed within areas where his forces were operational. Conversations overheard during his captivity indicate that Guderian thought that National Socialism was a fine idea misapplied.

ERWIN ROMMEL

1891–1944

MICHAEL BURLEIGH

IN THE 1951 FILM The Desert Fox, *Erwin Rommel was portrayed by the British actor James Mason as a noble knight in armour, disdainful of Nazi fanaticism. Ironically, Rommel was himself a bit of a film star, having figured prominently in Goebbels's film* Victory in the West, *and subsequently had one of the limping propaganda minister's experts permanently attached to his staff to keep his image well burnished. The real Rommel – the youngest German field marshal of the Second World War – was undoubtedly one of the most charismatic figures of that conflict, and even his most prominent opponents, including Churchill, Montgomery and Patton, became transfixed by him, to the point of obsession. Among the qualities of a great general that Rommel possessed was the ability to colonize the mind of his enemy – to the extent that Churchill grew tired of hearing his name.*

During the First World War Rommel had achieved a highly distinguished record as a combat commander, demonstrating enormous initiative and courage even after being wounded. Although he was the author of a book on infantry warfare – studied by General Patton among others – during the invasion of Poland in 1939 he quickly appreciated the importance of tanks, and went on to reveal as much skill in strategic retreat as in offensive operations. His mystique was further enhanced by his relative youth and by the nobility of his death in 1944 in the wake of a failed attempt to kill Hitler.

But how much of this stands up to critical scrutiny? Was it possible for any senior member of the German armed forces to insulate themselves from Nazi criminality? Did the war in the desert simply involve the British, Germans and Italians, with no reference to the surrounding Arabs? Despite the glamorous image of a man of action – the sunburned face, the dusty cracked lips, the crow's feet from screwing up his eyes in blinding light, the penchant for shooting gazelle with a sub-machine-gun from his staff car – Rommel was a prematurely sick man, suffering from depression, headaches, insomnia, low blood pressure, rheumatism and stomach trouble. A complex figure then, if not the matinée idol set up with the collusion of his former enemies.

The birth of a legend

Rommel was not a scion of the Prussian Junker caste used to the carmine stripe down their generals' trouser legs, but rather the son of two Swabian schoolteachers from a small town near Ulm in Württemberg. He was a short, shy boy, who just managed to pass the secondary-school leaving certificate or *Abitur*, without which it was impossible to become an officer in the German army. He joined the 124th Infantry Regiment as a cadet in 1910, and a year later was promoted from NCO to lieutenant. Rommel was a brave and physically tough soldier rather than a military bureaucrat. Upon the outbreak of the First World War he took part in the advance to the Marne, where he led his men into a fire-fight with French troops in the hamlet of Bleid. He was wounded in the thigh in September 1914 after taking on three French soldiers, despite having run out of ammunition. For this he received the Iron Cross second class. In early 1915 he stormed a French position and repulsed a counter-attack, for which he received the Iron Cross first class.

Rommel's future genius is partly attributable to the fact that he did not spend the entire war grimly stuck in the trenches of the Western Front. His war was more mobile and varied. Wounded again in the leg, Rommel was transferred to a newly formed mountain unit, which was deployed in the Vosges and then sent to Romania. In two separate operations, he captured 400 prisoners, moving on to take heavily defended enemy lines, while personally sustaining a wound in the arm. Transferred to the Isonzo front, he played a key role in the Austrian victory at Caporetto (1917). Constantly active for fifty hours non-stop as he took Monte Matajur with a single battalion, Rommel captured a total of 150 Italian officers, 9,000 soldiers and 81 artillery pieces. For this he received the *Pour le Mérite* decoration and promotion to captain. His next exploit involved swimming the River Piave with six men in order to surprise an Italian garrison in the village of Longarone. The garrison duly surrendered.

In view of this remarkable war record, Rommel was selected as one of the 4,000 German officers kept on after the Treaty of Versailles in the drastically scaled-down Reichswehr. By October 1929 he was an instructor at the infantry school in Dresden, where a book based on his combat experiences became a Weimar bestseller. By 1937 he was a full colonel, director of the war college at Wiener-Neustadt, the army's liaison officer to the Hitler Youth and from time to time head of Hitler's military security battalion. In August 1939 he became a major-general, attached to Hitler's field headquarters and in charge of the dictator's safety. This should give us pause regarding Rommel's alleged anti-Nazi sympathies, which did not extend to the miracle-working Führer. For although Rommel had reservations about

Nazism, this was counter-balanced by admiration for the dynamic dictator, especially since the latter had evaded the military restrictions of Versailles and rebuilt the armed forces as a pillar of restored German might. Hitler admired Rommel in turn as a brave combat soldier and a man whose background was almost as modest as Hitler's own.

After participating in the Polish campaign, in February 1940 Rommel was appointed commander of the 7th Panzer Division based at Godesberg. After rapidly mastering the techniques of armoured warfare, Rommel commanded his men in battle during the invasion of France, always leading from the front regardless of the dangers, and receiving two wounds in the process. Rommel's forte was speed and surprise, so much so that his division was nicknamed the 'Ghost Division': moving at 20–30 miles per hour, his troops captured tens of thousands of prisoners and took Lille within a few days. After an enforced rest and refit, his tanks crossed the Seine near Rouen and pushed the enemy back to Cherbourg, where 30,000 French troops followed Marshal Pétain's order to surrender. In six weeks, Rommel's armoured formation had taken over 100,000 prisoners as well as 450 enemy tanks, in return for modest losses.

Lord of the desert

While the RAF and the Luftwaffe contested the skies over Britain in the late summer of 1940, Hitler's partner in crime, the Italian dictator Benito Mussolini, set about realizing his dream of recreating the Roman Empire in the Mediterranean by attacking the British in Egypt, and later by invading Albania, Yugoslavia and Greece. On 13 September 1940 the Italian army advanced from Libya into Egypt in order to capture the Suez Canal, a vital artery for imperial British interests. Five poorly equipped Italian divisions under General Mario Berti forced the British to retreat until, reinforced, the latter counter-attacked. When the Italians fell back into Tripolitania (northwestern coastal Libya), Hitler rushed in his 5th Light Division to thwart the British; this was the advance guard of the legendary Afrika Korps, the formation that Rommel commanded in a theatre where the Italians were notionally in ultimate charge. Rommel's orders were to prevent the British from expelling the Italians from North Africa entirely. He arrived in the middle of a rout: the British had taken several major centres, including Tobruk, and much of Cyrenaica (eastern Libya), together with 130,000 prisoners. An attack on Tripoli was only averted because General Wavell had to divert a corps to participate in the Allied effort to eject the Germans from Greece.

Rommel's limited remit, established at a meeting with Hitler in March 1941, was to recapture Cyrenaica, with a force that was predominantly Italian. Despite the modesty of this initial aim, Rommel went on to turn North Africa into a theatre that occupied most of the energies of the British. They in turn became obsessed with their wily opponent, nicknaming him 'the Desert Fox'. Without waiting for adequate supplies, Rommel attacked at the end of March, capturing El Agheila, Benghazi, and all of Cyrenaica except Tobruk itself.

Desert warfare had certain unique features. Compared with the vast forces engaged on the Eastern Front, both Allied and Axis armies in North Africa were modest in size. The harsh terrain had its own problems, notably the constant need for water for both men and machines, while the few roads were long and dusty. It is usually claimed that the absence of the SS in this theatre meant that with some exceptions the war was conducted according to the Geneva Conventions, under the curious gaze of the indigenous population. That was not so. The SS *were* present, in the form of the mass murderer SS-Obersturmführer Walther Rauff, a gas expert, and some 2,500 Tunisian Jews died in camps over a six-month period. Moreover, at a political level, the Nazis were heavily involved with Arab nationalists and Islamists, notably the Palestinian Grand Mufti of Jerusalem, whose anti-Semitic exterminatory zeal matched that of Hitler. Had Rommel succeeded in taking the Afrika Korps through Egypt into British-mandated Palestine, it is not difficult to imagine the fate of the 80,000 Jews settled there, since even without such a campaign

Jewish population centres were subject to Luftwaffe bombing.

Germany's youngest field marshal

The war in North Africa moved to and fro over vast distances. In late 1941 and early 1942 Claude Auchinleck's Operation Crusader forced Rommel to retreat from Benghazi to Mersa el Brega, where he waited on developments elsewhere that might improve his fortunes. After Field Marshal Albert Kesselring, commander-in-chief of German forces in the Mediterranean theatre, had achieved air superiority off Italy, Rommel's supply lines were more secure: convoys could now cross the Mediterranean carrying heavy armour and anti-tank guns. In January 1942 Rommel launched an attack along the coastal road from El Agheila, taking Benghazi, Derna and the western half of Cyrenaica. While the British geared themselves up for a counter-attack, Rommel struck first with Operation Venezia, destroying the Free French at Bir Hacheim, and then hit the British Eighth Army at Tobruk. The latter lost 260 tanks in one engagement, and 30,000 men were taken prisoner. Hitler promoted the 49-year-old Rommel to field marshal, the youngest in the German army. Instead of pausing to recuperate and waiting for the Axis to seize Malta so as further to secure his supply routes against British air assault, Rommel felt emboldened to push into Egypt in pursuit of the wild dreams of linking up with German forces in southern Russia, or even of conquering India.

These dreams encountered reality at a minor railway stop in the Qattara Depression called El Alamein. There, in July 1942, Rommel's tanks were routed by Auchinleck's guns. Using 'Ultra' intelligence gleaned from Bletchley Park, where teams of British code-breakers had succeeded in unscrambling the German Enigma cipher, Auchinleck directed his counter-attacks at the Afrika Korps's Italian elements, which were thought less resilient than the German units. After an attack at Alam Halfa failed, Rommel dug the Afrika Korps in, anticipating a counter-attack by the man who in August 1942 had replaced Auchinleck as the new commander of the Eighth Army, Lieutenant General Bernard Montgomery. Wracked with illness and despairing about his extended supply lines, his reliance on captured enemy equipment and his Italian allies, Rommel went home on convalescent leave in September 1942. His efforts to persuade Hitler and Mussolini of the gravity of the situation in North Africa came to nought. He began to intimate to his wife his disillusionment with Hitler.

The 'panzer graveyard'

On 23 October Montgomery and General Alexander, the new commander-in-chief Middle East, launched the Second

Battle of El Alamein, fortified by the addition of 40,000 extra troops from Australia, India and New Zealand and 300 US Sherman tanks. In all, 200,000 men and 1,000 tanks moved against the Afrika Korps's 100,000 troops and 500 tanks. When Rommel's replacement, General Georg Stumme, died of a heart attack on 24 October, Hitler telephoned Rommel to urge him to resume his earlier command. Rommel anxiously looked on as the Battle of Stalingrad unfolded, aware that only a German victory on the Volga would free sufficient manpower to mount an invasion of Persia, forcing the British to call off the Eighth Army's attack at El Alamein by redeploying their main forces there. His gloom deepened as Allied aircraft succeeded in sinking one after another of his precious supply ships.

On returning to North Africa on 25 October, Rommel found chaos, with his men on half rations and only enough fuel for three more days. The Italian army had been bled white at Stalingrad, and its troops in North Africa were exhausted. They sometimes fought bravely, despite being equipped with rifles of 1890s vintage and grenades that were often lethal to the user; some of the simpler souls wore ribbons on their sleeves to help them distinguish left from right. By the time the Second Battle of El Alamein had come to a close on 5 November, virtually all of Rommel's armour had been wiped out; the battle became known as the 'panzer graveyard'. Hitler had ordered Rommel to 'stand fast'. Instead, Rommel ordered withdrawal.

From El Alamein to the Kasserine Pass

Rommel pulled out of western Egypt in early November 1942 just as US, British and Free French forces landed 1,400 miles to his west in French Morocco and Algeria. Rommel went on to conduct a brilliant fighting retreat to Mersa el Brega. His new orders were to defend Libya at all costs, prior to expelling the Allies from Tunisia. He flew to Hitler's Rastenburg headquarters to remonstrate that this was pointless and that the North African campaign was effectively over. A furious Hitler ignored him.

Back in North Africa, Rommel extricated himself from Libya with a view to linking up with Italo-German forces under Colonel General Jürgen von Arnim installed in a bridgehead in northern Tunisia. Arnim set about securing the passes through the Dorsale mountains in order to effect a junction with Rommel's more southerly forces. This set the scene for what became known as the Battle of the Kasserine Pass, which commenced in January 1943.

Rommel completed moving his Panzerarmee from Libya into Tunisia, where he was to be replaced by the Italian commander General Giovani Messe. With Montgomery paused at the Mareth Line just over the Libyan–Tunisian border, Arnim and Rommel were encouraged by

Kesselring to strike in a pincer movement through the Kasserine Pass and on towards the Allies' main supply dump in eastern Algeria. This would kill the Allied presence in Tunisia. Partly because of a lack of coordination with Arnim, in late February Rommel called off his attacks amidst the carnage of the Kasserine Pass.

Although Rommel had given the relatively inexperienced and poorly led Americans a nasty shock, the wider goals of the battle were unrealized. Rommel was briefly appointed commander of the entire Army Group Africa, but this was too late to halt the momentum of Montgomery's Eighth Army. After vainly pleading with Hitler to wind up the entire campaign, Rommel had to watch from the sidelines on sick leave as Montgomery destroyed the Axis forces in Tunisia, taking nearly a quarter of a million prisoners in operations that had a similar effect on Allied morale as Stalingrad.

Defender of Fortress Europe

As the Axis forces in North Africa faced final defeat in May 1943, Rommel was posted to a secret planning group to deal with the contingency of Italy dropping out of the war. After the Allied landings in Sicily in July, Rommel was tasked with the occupation of key nodal points in northern Italy to cover Kesselring's withdrawal from the south. From September 1943 onwards Rommel was based at a headquarters on Lake Garda. He and Kesselring differed as to whether Italy should be defended south of Rome or from Rommel's northern bastion. Hitler supported the overly optimistic Kesselring and appointed him supreme commander in Italy, while assigning Rommel to inspect the coastal defences on the Atlantic in anticipation of further seaborne invasions.

Although Gerd von Rundstedt was the western theatre commander, Rommel was the de facto commander of German forces in the west. In early 1944 he desperately tried to improve the Normandy defences, adding 5 or 6 million mines, together with concrete gun emplacements and clusters of 'Rommel asparagus', high poles designed to impede gliders, while the coastal shallows were covered with underwater obstacles and booby-trap devices. Rundstedt paradoxically became the main advocate of mobile warfare, holding fire-fighting forces in reserve to hit the Allies after they had got a toehold on shore; in contrast, Rommel wanted to spread his forces along the coast so as to massacre the Allies while they were up to their waists in sea water or pinned down on the open beaches. This made greater sense, as total Allied air supremacy meant that all German movements were lethally interdicted. With one field marshal distinguished by age and authority and the other by youthful charisma, Hitler found it impossible to go for one strategy or the other, so the German forces in France ended up labouring under a synthesis of both. Rommel was at home in Germany when, on the morning of 6 June 1944, the Allied invasion of Normandy started. It quickly became evident to him that the Allies were too strong to be dislodged and that it would be best to write off Normandy so as to defend their likely routes into Germany. Hitler refused to countenance this and urged Rundstedt and Rommel to counter-attack. They duly did so, but achieved little apart from losing more men and matériel. Hitler then replaced Rundstedt with Kluge, who rapidly came to the same gloomy conclusions as Rommel.

Death of the Desert Fox

On 17 July 1944 Rommel was badly wounded when two Allied planes attacked his staff car, leaving him with a skull fracture and other injuries. It was while he was recovering from this incident that his name was dropped into the vengeful atmosphere immediately following the 20 July attempt on Hitler's life: Rommel had been spoken of as a substitute head of government who would immediately broker a peace deal with the Western Allies.

While recuperating at home, Rommel received a visit from two generals, who presented him with a stark choice: face a treason trial, or commit suicide. If he killed himself, his wife and son would be left unharmed – a serious threat since the Nazis were murdering the families of many of the plotters. Rommel got into a car with the generals and swallowed a cyanide capsule; his body was incinerated to prevent a post-mortem. With characteristic cynicism, Hitler decreed that the young field marshal, only 53 years old when he died, should be given a state funeral, and insisted that an imposing memorial should be erected over Rommel's grave.

BERNARD MONTGOMERY
1887–1976

ALISTAIR HORNE

FIELD MARSHAL MONTGOMERY has been hailed as the greatest British field commander since Wellington, not a title for which there can be much competition. On the other hand, various details of his career, and particularly his personality, have rendered him one of the most controversial leaders of the Second World War. Over the years his reputation has ebbed and flowed, suffering most of all in the United States – as much through media-oriented criticism as from serious military studies.

Bernard Law Montgomery ('Monty' to detractors and fans alike) was born on 17 November 1887, into a Victorian-Irish family of slender means. His father was a churchman who became Bishop of Tasmania shortly after Monty's birth. His mother, Maud, seems to have been a harsh woman;

Monty grew up with little affection from her. He decided on an army career, claiming that he did it to annoy his mother. He could not afford his first choice, the Indian Army, joining instead the less 'posh' Warwickshire Regiment.

When the First World War broke out, Monty was 26, a full lieutenant. In action near Ypres he was shot through the lung by a sniper. Narrowly surviving, he was left short of breath for the rest of his life, hating tobacco smoke. He was awarded the DSO (Distinguished Service Order) and spent the remainder of the war on various staffs. He never forgot what he saw of the incompetence of the British army leaders and their wasteful expenditure of soldiers' lives. Monty ended the war a lieutenant colonel, a rank in which he stuck for two deadening decades. Among the polo-playing officer corps with private means, he came across as a boring misfit, with his single-minded professionalism and pursuit of higher standards. He became friends with another brilliant 'misfit' – the military historian and strategic thinker, Basil Liddell Hart.

In the 'wilderness years' of the 1920s and 1930s, Monty was described as having the 'self-abnegation of a monk'. In 1927 love struck the 40-year-old bachelor. He married Betty Carver; they had one son, David. Betty opened a wider world of culture, affection and fun to Monty, with a softening effect on his austere personality. Then tragedy struck; in 1937, an insect bite turned to septicaemia and Betty died. Monty was inconsolable. With war on the horizon, he once more devoted himself to soldiering, more single-mindedly than ever before. He was now a 50-year-old brigadier, on the verge of retirement. Then, in April 1939, after serving as commander of the 8th Division in Palestine for a year, he was summoned home to take command of the 3rd (Infantry) Division.

The early years of the war

Nicknamed the 'Iron Division', the 3rd was one of the handful of elite British units sent to France in the British Expeditionary Force (BEF) at the outbreak of war. The corps commander was Monty's fellow Irishman and long-time supporter, Lieutenant General Alan Brooke. Monty's self-confidence was immeasurable; he was going to command the most effective division – and he was going to win the war. But the BEF proved dismally unfitted for modern warfare. Monty brought the Iron Division through the retreat from Dunkirk in excellent fighting order. Back in England, it became the only division with sufficient heavy arms to equip it.

Monty was put in charge of the new V Corps, defending a denuded south of England, and was promoted lieutenant general. His first priority was fitness, 'physical and mental'. Officers had to be 'full of binge'; meaning that 'they must look forward to a good fight'. When inspecting troops, he asked them to remove their helmets so that he could see whether or not their eyes showed 'binge', the light of battle. Woe betide the commanding officer if they did not. To a stout colonel whose doctor had warned that an early-morning run would kill him, he observed that he should get it over with, so that he 'could be replaced easily and smoothly'. This robust approach gained Monty the respect of rank-and-file and junior commanders, but it tended to alienate his peers. When command of the battered Eighth Army in North Africa came up in August 1942, he was not Brooke's first choice.

Its morale thoroughly shaken, the Eighth Army had been steadily retreating before Rommel since January 1942, and now had its back to the Suez Canal at El Alamein, little more than 100 miles away. Monty was to take over command under General Alexander. The British army as a whole had hardly won one encounter with Hitler's Wehrmacht. Its equipment was inferior, and Eighth Army armoured tactics were not much better: during one terrible Balaklava-style charge in July 1942, the British had lost 118 tanks to three of the enemy.

Churchill persuaded Roosevelt to rush 300 Sherman tanks round the Cape to replace the heavy losses inflicted by Rommel's panzers, and by late October Monty would have a powerful superiority in weaponry. But more important than matériel was turning around a beaten army. Within three months, Monty's sharp features, like those of an aggressive Jack Russell, were to become the most celebrated in the Western world – to the amazement of his colleagues. He imposed clarity in his battle orders. He built up a massive weight of artillery and – for the first time on the Allied conduct of the war – moved the tactical airforce HQ alongside him. Above all he imbued the whole army, down to the lowliest private, with a sense of 'binge', that they were going to 'hit the enemy for six out of Africa'.

With his stonewall defence at Alam Halfa (30 August – 5 September), Monty blunted Rommel's final bid for Cairo and the Suez Canal. Yet the costly thirteen-day 'roughhouse' of Alamein itself (23 October – 5 November), the heavy casualties breaking through the German minefields, were the antithesis of everything Monty stood for. It was 'an untidy battle', but it brought the narrow margin of success. After Alamein, though, Monty's armoured *corps-de-chasse* failed to pursue the success cohesively: an over-cautious pursuit to Tripoli permitted the bulk of Rommel's army to escape. For this failure Monty has been criticized, but as he had admitted in his diary as late as October: '... the training was not good and it was beginning to become clear to me that I would have to be very careful ... that formations and units were not given tasks which were likely to end in failure because of their low standard of training ... I must not be too ambitious in my demands.'

Monty's victory at El Alamein was the first turning point in the war. Churchill, however, recognized that it was only the 'end of the beginning'; compared with the titanic battles on the Russian front, it was but a sideshow. Now followed Sicily and the long slog up through Italy, where for a variety of reasons, Monty did not shine. But Alamein had put him on the map as the general who could win. At the end of 1943 he was set in charge of Operation Overlord, the operation to take the Allied armies into northern France in June 1944.

Preparations for Overlord

Monty's overall boss was to be General 'Ike' Eisenhower, now nominated Allied Supreme Commander. Ike, a staff officer, had never seen 'the face of battle'. The two did not start off well. At their first meeting, Monty (then still senior in rank) had barked at Ike for lighting up in one of his conferences. Ike, a chain-smoker with a low boiling point, never forgave Monty. Monty's first move was to scrap the existing plans for the invasion with brutal bluntness. He increased the original three-division front to five. Even this was to leave an uncomfortably small margin of elbow-room from which to punch out of the beachheads. What set the limit was the shortage of landing-craft, which were being held back for a subsequent invasion of Mediterranean

France, a dispersal of effort pressed upon Churchill by Washington.

An essential ingredient of Overlord was the brilliant deception plan – pushed by Monty – which persuaded German intelligence that the main invasion would take place across the Pas de Calais. It pinned down the powerful Fifteenth German Army, which could otherwise have intervened decisively in Normandy. Apart from overseeing every meticulous detail of the biggest landing operation in history, Monty was responsible for the relentless bombing which would impede German movement of reinforcements across France. Meanwhile, he tirelessly raised 'binge', visiting a division a day, seven a week. Monty got it over to every man of the nervously waiting troops that they would only go in with immense supporting fire, and that they would win.

More controversial was Monty's presentation of a map of 'phase lines'. To Monty, geographical lines were unimportant; what counted was the destruction of enemy forces within that area. But what would hang like a millstone round his neck was a boastful assumption that Caen, the critical hub on the east of the British/Canadian sector, would be taken on D-Day itself.

Overlord

On 6 June 1944, D-Day, the Allied invasion of Normandy, began (see also pp. 272–5). Five out of eight assault brigades were British and Canadian; two out of the three airborne divisions were American. Of the aircraft deployed that day, 6,080 were American and 5,510 were from the RAF or other Allied contingents; but of the naval force, only 16.5 per cent were American (because of the demands of the Pacific War). The Americans landing on Omaha Beach had the misfortune to encounter a newly arrived German division and suffered horrendous losses. On the British sector, a major success was the capture of Bayeux on D-Day itself. But the failure to secure Caen for another two months was to prove costly; but what, for Monty, proved most disastrous – certainly in the eyes of his US critics – was his boast that 'everything had gone according to plan'.

Montgomery set up his HQ only 3 miles from the German lines. This may have seemed rash but it was a symbol that D-Day had succeeded, the Allies were firmly established – and the war in effect won. Monty exercised a personal and hands-on style of command. His use of forward HQs put him supremely in touch with the fighting formations under him, but it left him out of touch with senior staffs back in England. With every setback, the murmurings of Montgomery's many enemies – British as well as American – rose unchecked, and Montgomery was never there to explain his strategy to Eisenhower.

The conquest of Normandy

Events, however, vindicated his strategy of drawing the panzers on to the British front on the east end of the line and wearing them down, preparatory for the American break-out from the west by Patton's fresh US Third Army (see p. 291–3). Eight out of Rommel's nine panzer divisions were pinned down by British and Canadian armies in the Caen sector. If the Germans could have shifted just one towards Bradley, who had undergone a costly battle to secure Saint-Lô, it would have made his break-out more difficult – perhaps even impossible before August. Holding the 'hinge' aggressively at Caen also prevented any likelihood of intervention from the German Fifteenth Army locked in the Pas de Calais.

Monty fought three hard battles around Caen. Heavy bombers destroyed the medieval city, but the rubble made it difficult for British armour to advance. When he tried to break through the third time, lack of elbow-room – the narrow width of the bridgehead – meant Monty had no room for the massed armour to deploy in the east – also led to failure with brutal tank losses as the thin-skinned Shermans came up against dug-in lines of German '88s. For a while, it seemed there might be a stalemate in Normandy. Under extreme pressure from both the US and the British press, Churchill came close to sacking Montgomery.

At the end of July, however, Patton swung relentlessly towards Paris, and the fleeing remnants of the German armies were smashed in the trap at Falaise. Montgomery has been criticized for not closing the jaws of the trap, allowing a large number of the enemy to escape (though without most of their heavy equipment). Patton claimed that he could have closed the gap from the south, but this is not sustained by his chief, Bradley: Patton's forces were simply not strong enough to take such a risk. Nevertheless, Falaise meant the end of the Normandy campaign. The Germans were reckoned to have lost 450,000 men including 210,000 PoWs, more than 20 generals, 1,500 tanks and 3,500 guns. Allied casualties had by no means been light: 209,672, of whom 36,976 had been killed (roughly in a ratio of two British and Canadian casualties to three American), a testimony to how hard the Wehrmacht had fought, despite almost total lack of air support.

In strategic terms, the Battle of Normandy was a decisive victory to rank with Stalingrad. And no one can deny the key role Montgomery played in it. As one of his severest critics, Eisenhower's chief of staff, Bedell Smith, was to remark of D-Day: 'I don't know if we could have done it without Monty. It was his sort of battle. Whatever they say about him, he got us there.'

The advance on Germany

On 25 August Paris was liberated; on 3 September Brussels. Abruptly, the whole tenor of the war changed. Now the falling-out between Eisenhower and Montgomery began in earnest. Both American and British armies felt that the Germans would collapse as they had in 1918. But the Combined Chiefs of Staff, at a level above Montgomery and Eisenhower, had no contingency plan for the next stage if the Germans did not collapse.

After D-Day, proportions in Allied forces had swiftly changed until by May 1945 the US predominance was to become of the order of three to one. It was inevitable that American public opinion would now insist on Eisenhower as overall land commander, the role Monty had held since Overlord. Great coalition leader though he was, Ike was no strategist. His plans were to advance all the Allied armies to the German frontier, then await opportunity. Montgomery thought this 'broad front' strategy represented a dangerous dispersal of effort. He besieged Eisenhower with his scheme of a 'single thrust' towards the Ruhr and then Berlin, spearheaded by the British armies in the north. Meanwhile Ike was under pressure from Patton for the go-ahead for a single thrust into central Germany across the Saar. History was to prove that Patton and Monty were both right, and wrong: logistically, as well as politically, there were not sufficient supplies even for one 'single thrust'.

Antwerp, Arnhem and the Battle of the Bulge

The disaster at Arnhem in mid September – for which Montgomery was held largely responsible – has been regarded as the key to the disappointing end of 1944. Possibly Eisenhower was at fault in permitting Monty to carry out this uncharacteristically risky operation – while at the same time letting Patton press on at Metz and the Saar far to the south. Yet, strategically, of far greater consequence was the earlier failure to clear the approaches to Antwerp, the largest (and still largely undamaged) port in northwest Europe, straddling the River Scheldt.

To use the port, the approaches on both the Belgian and Dutch sides of the Scheldt estuary would have to be cleared of the enemy. After a lightning advance, Major General 'Pip' Roberts's British 11th Armoured Division had captured most of Antwerp itself by 4 September. Then, running out of fuel and with no further orders, Roberts halted. He claimed later that he could easily have covered the vital 20 miles to secure the Scheldt approaches. For vital weeks, while Hitler's armies regrouped, the Allies were deprived of Europe's biggest port, at a time when most of their supplies were still arriving via the beaches of Normandy 300 miles to the rear. Moreover, the German Fifteenth Army was permitted to escape across the Scheldt, its evacuation across the 3-mile-wide estuary in the teeth of Allied air and sea supremacy a feat little less remarkable than that from Dunkirk in 1940. These troops would be in position when Montgomery launched his airborne attack on Arnhem on 17 September.

Amazingly, neither Monty nor Ike appreciated the significance of Antwerp. But Alan Brooke also appears to have been blind to it; so too was Churchill. The failure at Antwerp constitutes the single biggest error of the Northwest Europe campaign. Success would have made Arnhem unnecessary.

The gamble of Arnhem had much to do with Montgomery's determination to prove that his strategy of the 'narrow thrust' into Germany from the north was right. As an undertaking, Arnhem had many planning defects; not least, it was a fundamental error to land the British 1st Airborne so far from the vital bridge. But Ike should surely have ordered Monty, after the Liberation of Paris, to go flat out for Antwerp. As it was, the failure at Arnhem left the Allied line stretched northwards, inviting a German counter-thrust. There were not enough reserves available, and Hitler had obtained a breather to create two new panzer armies.

The 'Battle of the Bulge' in the winter of 1944/5, victory wrested out of defeat at appalling cost – 80,000 US casualties for the cost of 120,000 Germans – remains an epic of American heroism. It opened the door to the invasion of Germany. But could it have been avoided through a better-conceived strategy back in September 1944? Montgomery always thought so. It was, though, appallingly maladroit when, on 7 January 1945, he held a disastrous press conference in which he could not resist a note of 'I told you so'. He paid high tribute to the American troops and to Eisenhower, but his reference to 'the Bulge' as 'a most interesting little battle' was insufferable to US pride.

Bradley never forgave him, and Eisenhower claimed it caused him 'more distress and worry' than anything else in the entire war. For the remainder of the campaign, Eisenhower paid minimal attention to Montgomery's strategic pleas; the two armies largely did their own thing.

Assessment

The discord in the last months of the war provided a sad ending to the brilliant display of Anglo-American amity that preceded Overlord – and it was compounded by the post-war 'Battle of Memoirs' that flooded out on both sides of the Atlantic. Montgomery's, in which he persisted in his smug claims that everything had 'gone according to plan', was bitterly criticized in America. He deserved better. But Monty's claim to fame was not just having 'got us there' on D-Day. It lay also in orchestrating the decisive defeat of Hitler's armies in France. Perhaps a last word should go to Winston Churchill, the man who appointed Monty in the first place, who often found himself infuriated by his arrogant intractability – yet recognized his surpassing qualities of generalship. When, after the war, members of his entourage were passing snide comments on Monty, Churchill bit back. 'I know why you all hate him. You are jealous: he is better than you are. Ask yourselves these questions. What is a general for? Answer: to win battles. Did he win them without much slaughter? Yes. So what are you grumbling about?'

In summary, five factors diminished the brilliant coup of 'Overlord', prolonging the war into 1945: 1. Failure at Arnhem; 2. The dispersal of landing-craft to the Mediterranean for the Anvil landings in the south of France; 3. Failure to seize Antwerp and its approaches; 4. The 'Battle of the Bulge'; and 5. Eisenhower's strategy to 'Bull ahead on all fronts' in 1945. Only the first came under Montgomery's jurisdiction.

GEORGI ZHUKOV

1896–1974

SIMON SEBAG MONTEFIORE

MARSHAL GEORGI KONSTANTINOVICH ZHUKOV was probably the greatest commander of the Second World War. Certainly, he played a leading role in all the decisive battles of the Eastern Front that decided the fate of the entire global conflict. He was either at the forefront of the planning, or in command, directly or indirectly, of the battles of Leningrad, Moscow, Stalingrad, Kursk, Operation Bagration, and Berlin – an astonishing record.

In the West, we celebrate the cults of British and American generals such as Montgomery and MacArthur, Bradley and Patton, while paying scant attention to the decisive Eastern Front. Zhukov is often ignored even though his achievements tower over our own military heroes in terms both of numerical scale and decisive importance. We should not overlook the much larger contribution of the Russian Front where the bulk of Hitler's army was ultimately destroyed. What is more, Zhukov is unusual in being the only Allied general to win victories against both the Japanese and the Germans.

Zhukov – who became the favourite general and Deputy Supreme Commander of the Soviet dictator, Josef Stalin – personifies the ruthlessness, brutality and crudity of the Stalinist system but also symbolizes the incredible courage and colossal sacrifices of the Russian people, who lost 27 million dead in what they call the Great Patriotic War. His triumphs and his failures always came at a terrible cost in casualties – indeed he prided himself on his ruthlessness. Arrogant, harsh, merciless, self-promoting, vain and unsubtle, he presided over costly disasters as well as remarkable victories. But he was neither sadistic nor devious but courageous, indefatigably energetic and drivingly optimistic in his campaigns against the Germans and the Japanese, often fearless in his dealings with Stalin, and brilliantly gifted as a battlefield commander.

From shoemaker's son to Soviet soldier

Zhukov was born on 1 December 1896, the son of a shoemaker peasant in Kaluga province, 124 miles south of Moscow. At 11, he was apprenticed to an uncle as a submaster furrier but in 1915, during the First World War, he was conscripted by the Tsarist army and remained a soldier for the rest of his life. In 1916 he was wounded, awarded two crosses of St George and promoted to NCO. In 1918 he volunteered to fight for the new Bolshevik government in the Civil War and served in the 1st Moscow Cavalry Division, fighting the Whites in the Urals and in the continuing battle for Tsaritsyn (later renamed Stalingrad), becoming a member of the Communist Party on 1 March 1919 and remaining in the Red Army after the Communist victory.

By 1931, aged 34, he was already known for his harsh competence and plain-spokenness. He was promoted to assistant to the legendary Inspector of Cavalry, Cossack commander in the Civil War, and Stalin's favourite general, Semyon Budyonny, who became Zhukov's protector and patron. Fortunately, he was not yet a senior officer when Stalin unleashed his savage purge of the Red Army officer corps in 1937, during which the tyrant shot three of the five Marshals of the Soviet Union and as many as 40,000 of his officer corps, devastating his command structure. Zhukov was vulnerable because of his rude and direct manner, his severity to seniors and subordinates alike. He was interrogated but ultimately his defiance – and the protection of Budyonny – saved him: he emerged from the bloodbath still tainted by the accusations against him but also as a respected young corps commander in a decimated army.

Japanese threat

In June 1939 Zhukov was summoned by Marshal Klim Voroshilov, People's Commissar for Defence, Stalin's top political general, and, like Marshal Budyonny, a crony from the Battle of Tsaritsyn in the Civil War, Stalin's baptism of fire as a warlord. This notoriously ignorant and bungling mediocrity appointed him to command Soviet and Mongolian forces that were facing a Japanese incursion close to the Khalkin Gol River. Defying and pushing aside the Stalinist political cronies, and taking firm command, Zhukov demanded reinforcements, which he duly received. He then launched a costly but powerful offensive against the Japanese, winning a decisive victory. Around 60,000 Japanese were killed or captured while Zhukov lost around 18,500; the combination of high cost, forceful offensive and ultimate victory remained Zhukov's trademarks. The Japanese defeat undoubtedly discouraged them from intervening against the USSR in 1941 and thus in its way was a decisive engagement. Had the Japanese been emboldened to intervene in 1941, Moscow would probably have been lost – and Russia may have fallen.

Summoned by Stalin, who immediately liked (as much as he liked anyone) and respected (as much as he respected anyone) the plain-speaking general, Zhukov was promoted to General of the Army and Commander of Kiev District. There, he commanded the annexation of Bessarabia in the summer of 1940, one of the gains of Stalin's 1939 Molotov–Ribbentrop Pact with Hitler.

German threat

At the end of 1940 Zhukov distinguished himself in the tense war-games ordered by Stalin that revealed the embarrassing ineptitude of the Soviet High Command. A furious Stalin sacked his chief of staff Meretskov and appointed Zhukov, aged 45, to the post on 1 February 1941. As Stalin rejected the convincing and repeated intelligence warnings that Hitler was about to attack, Zhukov, always aware how close he had come to liquidation in 1937, cautiously and uncomfortably tried to push the dictator towards precautionary readiness, despite the threats of Stalin's vicious henchmen. He urged vigilance and mobilization but only so far: Stalin's 'dungeons', he later admitted, were never far from his thoughts.

When Hitler launched Operation Barbarossa against the USSR on 22 June 1941 (see also pp. 199–203), it was Zhukov who rang Stalin at his residence to give him the news. When he finally got through and the dictator came to the phone, Zhukov said the Germans were attacking. He could hear Stalin breathing. 'Did you understand me?' repeated Zhukov. Finally Stalin summoned commanders and Politburo to the Kremlin.

In the chaotic routs and retreat towards Moscow that

followed, a floundering Stalin deployed the severe and blunt Zhukov as troubleshooter, first to the collapsing south-western front. Zhukov managed to lead a counter-attack, using his favoured Stalinist methods: 'Arrest immediately!' he would say concerning any retreating officers. 'Bring them to trial urgently as traitors and cowards.' Stalin always appreciated Zhukov's talent and brutal honesty; as Marshal Timoshenko recalled: 'You know Zhukov was the only person who feared no one. He was not afraid of Stalin.'

Zhukov had repeatedly warned Stalin of the dangers of Nazi aggression. Now, as the Soviet armies collapsed and disintegrated, facing encirclement and retreat, Stalin recalled him to Moscow.

Outspoken general

On 28 June Minsk fell, opening the road to Smolensk and Moscow. Stalin realized that the USSR was in danger of defeat. When he visited the Defence Commissariat to confront the generals, his toughest commander, Zhukov, admitted he had lost control of the front but rudely demanded the right to get on with his work. A row broke out between Zhukov and Beria, Stalin's fearsome secret police boss. 'Excuse my outspokenness,' Zhukov said to Stalin. But the dictator lost his temper, shouting at Zhukov, who burst into tears, being comforted by Stalin's chief henchman, Molotov. Stalin retired exhausted and demoralized to his mansion but emerged reinvigorated three days later to command the war as Supreme Commander-in-Chief.

Zhukov was henceforth his chief military adviser as the southern fronts now shattered. On 29 July Zhukov recommended that Stalin avoid encirclement of more troops by abandoning Kiev. 'Why talk rubbish?' shouted Stalin, at which Zhukov lost his temper: 'If you think the chief of staff talks rubbish, then I request you relieve me of my post and send me to the front,' shouted Zhukov. Stalin was shocked – and impressed: 'Don't get heated,' he answered, 'but since you mention it, we'll get by without you. Calm down, calm down.' Zhukov was sacked as chief of staff – but Stalin kept him as a member of Headquarters, the Stavka, and almost uniquely admitted later that Zhukov had been right.

Leningrad and Moscow

Meanwhile Leningrad was in danger of falling to the Germans: on 8 September 1941 Stalin dispatched Zhukov to save the city and remove the bungler Voroshilov. Displaying sang-froid, merciless Stalinist discipline (including the death penalty used liberally) and military skill, Zhukov stabilized the front, even counter-attacking. Hitler cancelled his assault and decided to starve, not storm, Leningrad into submission.

On 5 October Stalin summoned Zhukov back to Moscow, which was increasingly in danger of falling to the Germans. Stalin was considering abandoning the capital, which would probably have heralded Soviet defeat and therefore Nazi hegemony over Europe. Had Moscow fallen, the entire history of the world would be different. It was the biggest battle of the entire war, indeed the largest in human history. Seven million men fought and 926,000 Soviet soldiers died (more than combined British and American casualties in the entire war). In all, 2.5 million were killed or wounded, 2 million of them Russian. This was Zhukov's battle.

Leaving Leningrad to its siege in the autumn of 1941, Hitler switched his panzers to Operation Typhoon, the taking of Moscow. Battle started on 30 September. On 3 October the panzer general Guderian took Orel and the Soviet fronts collapsed. Panicked, Stalin called Zhukov in Leningrad: 'I've got just one request. Can you get a plane and come to Moscow?' 'I'll be there,' said Zhukov. On 7 October Stalin received Zhukov and ordered him to save Moscow. As the fronts on all sides of Moscow were collapsing, Zhukov had only 90,000 men. He nonetheless took control of the colossal conflict. Visitors to Stalin's office were amazed by Zhukov's 'commanding tones as if he was the superior officer, and Stalin accepted this'. But the advance continued. Law broke down on the Moscow streets. By 15 October Stalin was considering whether to abandon the city. On the 16th, ministries and all the embassies were dispatched to Kuybishev in the rear. But still Stalin refused to leave, demanding on the 17th that order be restored. By 18 October Kalinin in the north and Kaluga in the south had fallen and there were panzers on the battlefield of Borodino. On the 19th Stalin declared a state of siege and savage punishments for any defeatism or panic as Zhukov commanded the desperate struggle with dwindling reserves. In Berlin, the Reich Press Office declared 'Russia is finished'. But the Germans – convinced victory was theirs – rested a day as the temperatures sank to a deep freeze. Stalin, now convinced that Japan would not attack Russia, gave Zhukov his hidden reserve: the crack 700,000 troops of the Far Eastern Army. On 7 November Stalin showed his defiance by holding the annual parade in Red Square. On the 13th, Zhukov reluctantly agreed to Stalin's insistence on a counter-attack but it wasted more troops and was subsumed in the last German push of 15 November. Stalin called Zhukov to ask: 'Tell me honestly as a Communist, can we hold Moscow?' 'We'll hold it without a doubt,' said Zhukov. On 5 December, having lost 155,000 men in twenty days, Zhukov fought the Germans to a standstill. On 6 December Stalin gave Zhukov three new armies. Zhukov planned a counter-offensive on the four nearest fronts. He managed to push the Germans back 200 miles. Moscow was saved. Zhukov was so exhausted that he slept for many hours. Even when Stalin called, his adjutants could not wake him. 'Don't wake him up until he wakes himself,' said Stalin, 'Let him sleep.' It was the first Soviet victory but a very limited one. German forces were not destroyed yet and could fight another day. But for now, Zhukov had saved Russia.

In the following months, Zhukov tried to restrain Stalin's craving for vast offensives and his lack of military comprehension that lost entire armies in German encirclements and cost many millions of Soviet troops. In May 1942 the Kharkov offensive ended in disaster, opening up the road to the city of Stalingrad and the Caucasian oilfields that Hitler needed to fuel his exhausted war machine.

Stalingrad

The Battle of Stalingrad became a savage struggle to save the city, but gradually Hitler's Sixth Army was sucked deeper and deeper into the ruined city until it became vulnerable to Soviet counter-attack and encirclement. On 27 August 1942 Stalin promoted Zhukov to Deputy Supremo. Zhukov refused: 'My character wouldn't let us work together.' Stalin replied: 'What of our characters? Let's subordinate them to the interests of the Motherland.'

On 12 September Stalin called Zhukov and chief of staff Marshal Vasilevsky to the Kremlin. It was a historic meeting. All three stared at the map. 'There might be another solution,' muttered the generals. 'What other solution?' asked Stalin. Before they could answer, he ordered them to work out a plan. The three of them evolved the concept of Operation Uranus, the encirclement of the German Sixth Army. The next night, Stalin greeted the generals uniquely with a handshake and agreed the plan with the words: 'No one else knows what we three have discussed here.' Stalin, after losing 6 million men through his own follies, had finally started to take military advice and become a competent supreme commander. Zhukov and Vasilevsky worked out details of the plan, but just before it was launched Stalin appointed Zhukov to command a huge diversionary offensive, Operation Mars, on the Kalinin and western fronts further north. Operation Mars, with its brutal frontal assaults, was Zhukov's biggest and most costly defeat. He admitted later that, throughout the war, he sometimes cleared minefields or took positions with savage frontal assaults regardless of the human cost: he was very much a Stalinist general. Uranus became the decisive victory of Stalingrad, which began the defeat of Hitler.

On the offensive

In January 1943 Zhukov was promoted to Marshal of the Soviet Union. A relationship of rough, guarded but mutual respect had grown up between Stalin and Zhukov; over meals in the early hours, they sometimes discussed even personal matters such as Stalin's relationship with his mother and sons. Zhukov said Stalin was like a temperamental boxer who got excited and always wanted to give battle, even when he wasn't prepared.

On 5 July 1943 Hitler launched Operation Citadel – an attack of 900,000 men and 2,700 tanks on the vulnerable Soviet bulge at Kursk. Zhukov was at the forefront of the fortification of Kursk, which withstood the German attack, and then, on 12 July, of the launching of the Soviet offensive that became history's biggest ever tank battle and the defeat of Hitler's last big offensive on the Eastern Front. For the rest of 1943 – during the battle to retake the Ukraine, and during the colossal offensive in Belorussia, Operation Bagration, planned and overseen by Zhukov in 1944 – the Deputy Supremo remained the top Russian general as the Germans were chased off Soviet soil.

As Soviet forces halted outside Warsaw, Stalin retook personal command of the fronts from Deputy Supremo Zhukov, whom he now appointed to command of the 1st Belorussian Front. On 1 April 1945 Stalin ordered Zhukov and his rival, Marshal Konev, commander of the 1st Ukrainian Front, to race each other to take Berlin, whatever the costs. Zhukov was to take Berlin from the Oder bridgeheads over the Seelow Heights while Konev was to push through Dresden and Leipzig, with this northern flank thrusting towards southern Berlin. Between them they had 2.5 million men and 6,250 tanks.

On 16 April Zhukov attacked the Seelow Heights outside Berlin with 14,600 guns, but on being repulsed and under the pressure of Stalin's irritation, he stormed the Heights at a terrible cost of 30,000 men. Stalin taunted Zhukov and then allowed Konev to push towards Berlin. Both marshals fought their way into the city. On 1 May Zhukov called to inform Stalin that Hitler had committed suicide: 'So that's the end of the bastard,' said Stalin. 'Too bad we couldn't take him alive.' Hitler's body was found by the secret police and taken back to Moscow, but Stalin pretended to Zhukov that he did not know this. Now the war was over, Zhukov was no longer Stalin's confidant.

Hounded hero

Zhukov was now Russia's greatest military hero. He dispatched planes of booty and art back to Moscow. His mistresses were top Russian singers and he travelled with a huge entourage, but he rashly implied in a press conference that he was the chief architect of the Soviet victory. Stalin, who promoted himself to Generalissimo, was suspicious of Zhukov's fame and prestige. On 9 May 1945 Zhukov presided over the official German surrender, himself supervised by Stalin's henchman, Vyshinsky. When Zhukov met

US General Eisenhower, he shocked the American by explaining that his method of clearing minefields was to send infantry running across them.

On 24 June Stalin gave Zhukov the honour of reviewing Soviet forces on a white Arabian charger at the victory parade in Moscow. But Zhukov was now out of favour. The Western press actually cited him as a potential heir to Stalin and the dictator even probed this idea: 'I'm getting old,' he told Marshal Budyonny. 'What do you think of Zhukov as my successor?' The secret police were soon investigating Zhukov and raided his apartments to find a 'Museum' of war booty, art and furniture. In June 1946, at a meeting of the Supreme Military Council, Stalin orchestrated an attack on Zhukov, who was then demoted to command the minor Odessa Military District. His officers were arrested and tortured, his houses searched, his booty confiscated and he suffered a heart attack, though Stalin refused to order his arrest. When the secret police presented evidence of conspiracy against him, Stalin replied: 'I don't trust anyone who says Zhukov could do this. He's a straightforward, sharp person able to speak plainly to anyone but he'll never go against the Central Committee.' Politically he had to destroy Zhukov, but personally he respected him.

After Stalin

On Stalin's death on 5 March 1953, Zhukov was recalled to become Deputy Defence Minister. In June he assisted the rising Nikita Khrushchev in the arrest and later liquidation of Lavrenti Beria, the secret police chief and temporary Soviet strongman for three months after Stalin's death. Khrushchev became the all-powerful Soviet leader – though he was no Stalin, either in terms of statesmanship or cruelty. In 1955 Zhukov was appointed Defence Minister. In 1957 he joined the ruling Presidium (Politburo) and when Khrushchev was almost overthrown by Molotov and Stalin's old henchmen, Zhukov helped Khrushchev defeat them. But Khrushchev was, like Stalin before him, jealous of Zhukov's power and suspicious of his ambition. Soon afterwards that same year, he accused Zhukov of 'Bonapartism', and sacked him. Zhukov remained in retirement under his death in 1974.

A statue of Zhukov on horseback has recently been raised on Red Square in Moscow. He was truly the greatest military hero of the Soviet Union. His monstrous faults, remarkable gifts and colossal achievements earn him a special place in the pantheon of Russian heroes with Prince Alexander Nevsky of Novgorod, Prince Alexander Suvorov, who served Catherine the Great and Paul I, and Prince Mikhail Kutuzov, Alexander I's commander of 1812. In

1942, Stalin created the Orders of Suvorov and Kutuzov to reward courage and victory, medals that are still awarded in the Russian army today. One day, there will be an Order of Zhukov.

IVAN KONEV

1897–1973

RICHARD OVERY

THERE ARE AMONG THE RED ARMY COMMANDERS of the Second World War perhaps a dozen who are known in the West among those with an interest in military history, but as household names there are only two: Marshal Zhukov and Marshal Ivan Konev. Konev won his reputation in four years of gruelling combat on the Eastern Front. His capacity to hold together large forces and to manage a complex and fluid battlefield for the conquest of much of central Europe in 1944 and 1945 marks him as a commander of exceptional quality. He was not a strategic thinker, but under the duress of war he acquired a firm grasp of operational realities.

Konev, like any other senior Soviet commander, operated as part of a large team, dominated by supreme headquarters and the General Staff, and above all by Stalin. It is difficult for the historian to decide how much independence of action any Soviet commander had, but it was essential once a task had been set that the operational commander fulfilled what was required of him. Konev, like all other Soviet military leaders, was measured on performance in the field. Since the campaigns from 1943 were almost continuous, operational capacity was constantly tested. Survival of the fittest was built in to Red Army performance. This, too, says much about Konev's qualities as a military commander.

The early years

Ivan Stepanovich Konev was born on 28 December 1897. This is one of the few facts from his early life about which there is agreement. Konev is usually described as coming from a village near Podosinovsky in central Russia, but his records show that he was born in the village of Lodeino, near Nikolsk in northwestern Russia. His family were poor peasants, though a later attempt to blacken Konev's name in the 1930s produced evidence that his father had been a *kulak*, a well-to-do peasant, hiring labour and owning livestock. Konev had an elementary education, worked on the farmstead until he was 12 and then became a lumberjack before being called up into the Tsarist army in April 1916.

He immediately attracted attention and was sent on a course for NCOs. He had seen no fighting by the time he was posted to an artillery division, and when the Revolution came he returned to Nikolsk, where he became a local Bolshevik military commissar and a lifelong and committed communist. He helped to suppress the Kronstadt rebellion in 1921, staged by disgruntled sailors near Leningrad, and during the Russian Civil War became first commissar of an armoured train, then commissar of the 17th Maritime Corps in Nikolsk. He was posted to the Ukraine in 1924, and in 1926 attended staff courses at the newly founded Frunze Military Academy. He commanded the 17th Rifle Division from 1926 to 1932, and then attended further staff courses. When Stalin's purges started to sweep through the armed forces in 1937, Konev was fortunate to survive, since he had been serving under General Uborevich, one of those executed. He was posted in 1938 to command the 57th Special Corps in Mongolia, most of whose officers had already been purged, and rose rapidly thereafter. In May 1941 he was made commander of the Nineteenth Army in the Ukraine, which is where he was when Germany invaded on 22 June 1941.

Konev and the defence of Moscow

The next four months were the most difficult of Konev's career. He was unfortunate to be in command of an army right in the path of the German attack, exposed like other commanders to Stalin's wrath when the Soviet front caved in. But Konev's early battles marked him out as distinctive. Personally brave, a tough commander who fought along-side his men, Konev was among those who imposed tempo-rary reverses on the German attacker. His Nineteenth Army drove the Germans out of Vitebsk before being forced to retreat towards Smolensk. It was here that Konev's army took part in the Red Army offensive to try to dislodge the German hold on the town, though it failed. On 12 September Konev was promoted to head the Western Army Group of six armies to prepare for the defence of Moscow, but the losses sustained at Smolensk and the strengthening of the German Army Group Centre led to disasters around the small town of Vyazma, where his long front line was based. German armoured thrusts cut through the weak Soviet forces and linked up behind Vyazma, ensnaring several of Konev's armies, which were forced to surrender. German armies netted 690,000 prisoners during the operation.

The disaster was scarcely Konev's fault, since the strength of the two sides was so unbalanced, but he had also been exposed to a major operation after just two weeks in high command. Nor had Soviet intelligence understood the nature of the German attack. Stalin thought about punishing Konev, and an investigation was begun into the Vyazma disaster, but General Zhukov, according to his own account, successfully interceded with Stalin, took Konev on as his direct deputy for the battle in front of Moscow, and placed him as commander of the Kalinin Army Group northwest of the capital, with the new rank of colonel general. Here he was able to prove himself as a commander at last. He stabilized the northern part of the Moscow Front around Kalinin, absorbing the German attack before launching a counter-offensive on 5 December, along with the rest of the Moscow Front, which led to the recapture of Kalinin. In January Konev's army group was strengthened and he was ordered together with Zhukov to try to encircle the bulk of German Army Group Centre around the Vyazma–Smolensk axis; but although Konev's forces succeeded in penetrating German lines, the operation was too ambitious in mid winter with limited resources, and Smolensk remained in German hands.

Konev was fortunate to have survived the crisis of autumn 1941, but thereafter he remained one of the favoured inner circle of senior generals, commanding the Western Front in August 1942, the Northwestern Front in March 1943 and finally the reserve Steppe Front, created behind the Kursk salient in June 1943 to lead the planned counter-offensive after the German attack (Operation Citadel) had faltered. On 3 August 1943, directed personally by Zhukov, his army group, together with the Voronezh Army Group under General Vatutin, drove the German army rapidly back, liberating Belgorod and finally Kharkov, the Soviet Union's fourth largest city, which had changed hands between the two sides so often in early 1943 that the urban area was completely destroyed. The success of the Kursk counter-offensive opened the way for a Red Army drive across the Ukraine to the River Dnepr. Konev's army group was renamed the Second Ukrainian Army Group, one of four responsible for driving back the whole of the centre and south of the German front. Between August 1943 and April 1944, when Konev's army group entered Romanian territory at Botosani, his forces were in almost continuous combat, driving the retreating German army back across the Dnepr and then on to the frontiers of eastern Europe.

The battle for which Konev was made a marshal of the Soviet Union, and which he came to regard as a model operation, took place from 24 January to 17 February 1944 in the southern Ukraine, after Konev's Second Ukrainian Army Group had pushed German forces back from the River Dnepr around Dnepropetrovsk. German forces held on to their positions on the river further north around Korsun, creating a large salient that Konev tried to eliminate by a strategy of encirclement. The battle that followed led to the destruction of what became known as the 'Korsun–Shevchenkovsky pocket' (sometimes known as the Cherkassy pocket) and the elimination, according to Soviet estimates, of 77,000 men.

Although Konev later claimed that he had directed and won the battle around Korsun, it was, like most Soviet offensives, in part a joint effort. Marshal Zhukov, who was the Soviet headquarters representative in the Ukraine, recommended trying to pinch off the salient, along the same lines that had been used to encircle General Paulus at Stalingrad. South of Korsun, Konev's Second Ukrainian Army Group was to strike northeastwards behind the salient, while General Nikolai Vatutin's First Ukrainian Army Group struck southeast. The plan was to make a wide corridor defended on both sides against enemy counter-attacks while the pocket was reduced by sustained air and ground attack. Konev set up a complex deception plan to mislead German defenders about the true destination of his attack, although unusually it failed to work this time. The attack, launched on 18 January and fought in bitter weather and atrocious ground conditions, brought ten days of hard fighting before the two Soviet army groups met at the village of Zvenigorodka, trapping the German XI and XLII Army Corps, with six divisions (including the SS Wiking Division), inside the pocket.

The German Army Group South under Field Marshal Erich von Manstein was short of tanks and aircraft, but he attacked the whole Soviet circle, on Hitler's instructions, rather than trying to break through to the pocket at one particular juncture. Konev proved able to control a difficult and dispersed battlefield, and prevented any serious penetration of the line. By 8 February he felt confident enough to invite the German commander in the pocket, Lieutenant General Wilhelm Stemmermann, to surrender. He refused, and Stalin began to fear that a breakout was planned. On 12 February Konev was given unitary control of the whole operation, and once again a more concentrated German effort to reach the south side of the pocket was blunted by Konev's redeployments, and only got to within 6 miles of the trapped German divisions. But on 15 February Stemmermann was ordered to do anything to try to break out, and over the next two days an estimated 35,000 escaped by filtering through Soviet lines, leaving almost all their equipment behind.

Konev ordered attacks on the escaping garrison with tanks and cavalry, but only the rearguard was caught and slaughtered, including Stemmermann, whose body was found on the battlefield. Out of an original trapped force of a little under 60,000, around 19,000 were killed or captured (and not the 77,000 claimed), but this does not include the substantial casualties inflicted on Manstein's divisions trying to reach the pocket. The news of the escapes was not revealed to Stalin, but Konev's operation, in difficult fighting conditions against a desperate enemy, inflicted a major operational defeat, with heavy losses of men and equipment, and opened the way to the broad Soviet advance across the southern Ukraine that began a few weeks later.

Konev brought a number of operational priorities of his own in commanding enormous numbers of men on complex and fluid battlefields. He adopted the Red Army tactic of concealment and deception, and generally used it successfully. Time and again the German side was caught out by uncertainty as to Soviet intentions. Konev also placed a top priority on learning about the enemy's dispositions as closely as possible, using a number of forms of reconnaissance and relying on infiltration and scouting. He saw this as one of the principal aims of any divisional commander. The main object was to find out where the enemy artillery was concentrated so as to direct fire without waste and in a sudden devastating barrage. The short, sharp artillery attack was then followed by rapid movement forward before the enemy forces had time to recover or re-establish communication. Rather than large numbers of infantry, Konev later wrote that 'the decisive factor in combat is fire'. Konev also favoured the use of tanks, strongly supported by aircraft, which he sent forward in numerous small penetration operations, breaking through into the enemy rear and creating confusion, rather than massing them for one major attack. The effect was to create panic among the defending troops and speed their withdrawal or surrender. The very great numerical superiority enjoyed by the Red Army and Air Force by 1944 and 1945 made this use of tanks a realistic tactic, and it explains the alarm felt by German soldiers when they faced an enemy who seemed capable of moving forward remorselessly, like a swollen river.

Konev's greatest campaigns

The first of Konev's major offensives in 1944 was the operation launched on 13 July 1944 in western Ukraine, which brought his forces to the Carpathian Mountains and the River Vistula in Poland. He was now the commander of the First Ukrainian Army Group, following the death of General Vatutin; it was the largest single Soviet force, comprising seven infantry and three tank armies, and four mobile corps, a testament to his growing stature as an operational leader. His force occupied the southeastern part of Poland in little more than two weeks, reaching the Vistula by 29 July. The all-important rail city of L'vov was the major target, and Konev's forces not only encircled and captured the city on 27 July, but succeeded in surrounding a force of 40,000 Germans at Brody, in another classic encirclement operation. Konev's army group then rested after months of combat while the Red Army further south moved into Romania and Bulgaria.

The second major operation was the so-called Vistula–Oder operation. This once again paired Zhukov and Konev as army group commanders, after Stalin decided that Zhukov should personally command the operations into Germany. Konev and Zhukov both now led army groups vastly larger in size and equipment than the German enemy they faced. Nevertheless the last two major operations, to the River Oder and then to Berlin in May 1945, were carried out with great speed and completeness. Konev's army group began its operations on 12 January, and crossed most of Poland to reach the Oder and Silesia in only two weeks. Both Konev and Zhukov, who had also crossed Poland, wanted to press on further, and Konev moved another 75 miles nearer Berlin to the banks of the River Neisse. But determined German resistance and supply problems slowed down the whirlwind advance, and Konev and Zhukov had to wait until April before the high command in Moscow was satisfied that Berlin could be seized with certainty.

The 'race for Berlin' was perhaps Konev's most famous operation. In March 1945 Stalin ordered firm operational preparation for the capture of the German capital employing Konev's First Ukrainian and Zhukov's First Belorussian Army Groups as the major players. Plans and redeployment had to take place very quickly, but Konev succeeded in extricating his armies from the conquest of Upper Silesia and had them prepared for attack by mid April. On 16 April he and Zhukov began their attacks, Zhukov directly opposite Berlin, Konev to the south. Although Konev was supposed to use his force to push south and southwest,

Stalin deliberately left open the demarcation line between the two army groups in case Konev might be needed to help capture Berlin. As Zhukov's attack stalled, Konev's succeeded rapidly and on 17 April Stalin telephoned Konev to find another solution. Konev told him that he could easily swing his armour northwards towards the German capital, and Stalin gave him the go-ahead. By 21 April Konev's advance guard had reached the outskirts of Berlin, and by the 24th had made contact with Zhukov's armies, which had now moved forward as well. In the end Konev's forces held back while Zhukov took the glory of the capitulation of Berlin, but it was now evident that the apprentice Konev had become a major rival, as capable of mastering a vast battlefield and destroying enemy formations at speed as his one-time mentor. Konev had the distinction of overseeing the last fighting in Europe. His forces moved south as well as north in late April, reaching the Czech border. Here German resistance continued, and on 9 May Konev's forces occupied Prague; two days later he took the surrender of the remnants of General Schörner's Army Group Centre, three days after VE-day was celebrated in western Europe.

Konev's post-war career

On 5 May 1945 Konev was appointed commander-in-chief of the Central Group of Forces in Austria and part of

Hungary. In April 1946 he was made commander of land forces in the Soviet Union. Unlike many other leading Soviet commanders, Konev kept Stalin's confidence after the war, perhaps because of his mounting rivalry with Zhukov. When Stalin summoned Zhukov before the Central Committee in 1946 and accused him of conspiracy, self-glorification and corruption, Konev was present. Although he denied that Zhukov could have been disloyal, he did agree openly that he was a difficult and temperamental personality. Zhukov was demoted, while Konev took his place as Red Army commander-in-chief. It is tempting to see in Konev's behaviour mere political opportunism, but his relationship with Zhukov was difficult even during the war years, perhaps because Konev himself sensed that as an operational commander he had grown at least to be Zhukov's equal, but got little of the credit.

Konev, a popular and well-respected Red Army leader and the most important figure in the post-war military establishment, went on to become commander-in-chief of Warsaw Pact forces, and in this role oversaw the ruthless suppression of the Hungarian Revolution in 1956. Five years later he was posted to be commander of Soviet forces in East Germany during the crisis over the Berlin Wall. In 1963 he went into retirement, and eight years later died of cancer. As a twice-decorated Hero of the Soviet Union he was buried in the Kremlin Wall.

DWIGHT D. EISENHOWER

1890–1969

CARLO D'ESTE

DWIGHT EISENHOWER gained a place in military history on 6 June 1944, the day the Western Allies invaded Normandy and thereby changed the course of the Second World War. As Supreme Allied Commander, the responsibility for carrying out the campaign to defeat Germany was his and his alone. Eisenhower was called upon to make one of the most difficult and courageous decisions ever required of a military commander: to launch the Allied invasion of Normandy despite bad weather which threatened to wreck the invasion. Few commanders have ever faced such extraordinary responsibilities, and that decision in itself, with its awesome potential for failure, changed the course of history.

The third of the seven sons of David and Ida Eisenhower, members of a strict, pacifist Mennonite religious sect called the River Brethren, Dwight David Eisenhower was born in Denison, Texas, on 14 October 1890, during a violent thunderstorm. Shortly thereafter the family moved to Abilene, Kansas. Although the Eisenhowers were so impoverished that wearing shoes was deemed a luxury, Dwight, known by the nickname 'Ike', grew to manhood in a happy home. He and his brothers were taught the values of honesty, hard work, independence and responsibility from an early age.

Ike at West Point

Eisenhower never aspired to become a professional army officer but, too poor to advance beyond high school yet determined to earn a college degree and eventually escape poverty, he passed a competitive examination and entered the United States Military Academy at West Point in 1911. A popular but undistinguished cadet, Eisenhower never took West Point seriously. He never pushed himself to excel, managed to accumulate a high number of demerits, walked numerous punishment tours and was fortunate not to have been expelled for his frequent violations of regulations. An inveterate prankster, his antics and irresponsibility ranged from inattention to his studies, smoking (an expulsion offence), leaving West Point without permission (another expulsion offence), and high jinks such as pouring water from buckets on unsuspecting cadets in the barracks.

Ike was a superb athlete and a promising football player, but his knee was severely damaged early in his cadet days and further injured when he fell from a mean-spirited pony during a dangerous equestrian drill. He was very nearly not commissioned because of it. He also had back and stomach trouble and a host of other ailments that late in his life led to ileitis (an intestinal complaint) and heart disease. These health problems were exacerbated by an addiction to cigarettes begun at West Point, which by the Second World War had escalated to four or more packs a day.

His West Point class became the most famous in Academy history and was later known as 'The class the stars fell on'. During the Second World War it produced two five-star generals (Eisenhower and Omar N. Bradley) and more than a dozen division commanders, most of whom served under their classmate Eisenhower in the Mediterranean and European theatres.

Early military career

Eisenhower entered the army in 1915 a second lieutenant of infantry with few career aspirations and a happy-go-lucky attitude that soon changed into one of professional seriousness. In 1916 he married Mamie Doud, whom he had met shortly after arriving at his first duty station in San Antonio, Texas. Their remarkable marriage lasted until Ike's death in 1969.

Eisenhower made his mark as a superb trainer of troops, a duty he came to detest as one training assignment followed another during the months leading up to American participation in the Great War. He desperately wanted a combat assignment in France in Pershing's American Expeditionary Force (AEF). Instead, he was assigned to the new tank corps and sent to command Camp Colt at Gettysburg, Pennsylvania, where he trained men destined for service in the AEF tank corps commanded by a young officer named Patton. When the armistice ended the war on 11 November 1918, Ike was devastated at not having seen combat and perceived his career a failure. 'I will make up for this,' he declared in anger and frustration.

The inter-war years

In 1919, the age of the motor vehicle was in its infancy. Eisenhower was offered a chance to escape the routine of the peacetime army life by becoming a member of what was called the Transcontinental Motor Convoy. In an amazing feat, a US Army convoy of trucks and assorted vehicles travelled from Washington, DC to San Francisco in an epic seventy-nine-day journey across a virtually roadless America.

In the United States, the terrible bloodshed in Europe bred a revulsion for future war. By the early 1920s the great patriotism of the First World War had turned into a national aversion to all things military and to the symbol of war – the military establishment. A militant pacifism took seriously the notion that America had indeed fought 'the war to end all wars'. Common symbols of the times in the 1930s were signs proclaiming: 'Dogs and soldiers, stay off the grass.' The army consisted of only a little over 100,000 men. During those years Ike served in a series of staff assignments, all the while wishing for but not getting troop assignments, and believing his career would end in his retirement as a relatively junior officer.

He served sixteen years as a major, studied tank tactics with his friend George S. Patton, learned to fly, studied war with his mentor General Fox Conner, served on the American Battle Monuments Commission under General John J. Pershing, and in the 1930s toiled for nearly nine years as an aide and staff officer to General Douglas MacArthur in Washington and the Philippines.

Pearl Harbor to Overlord

With the advent of the Second World War, chief of staff General George Marshall purged the army of its many ageing officers. Those over 50 were unceremoniously retired. Two notable exceptions were Eisenhower and Patton, each of whom had become too valuable to retire on the grounds of age alone. In 1941, as America was being drawn into war with Japan, Italy and Nazi Germany, Eisenhower's aspirations were modest. He would have considered himself successful merely to serve as a colonel in an armoured division under the command of his long-time friend, the flamboyant Patton. Instead, his star rose dramatically that year as a result of the Louisiana Maneuvers – the largest peacetime military training exercises ever conducted in the United States. As the chief of staff of the Third Army, Eisenhower's brilliance as a planner was on full display and resulted in his promotion to brigadier general.

After Pearl Harbor (7 December 1941), he was summoned to Washington to become the army's chief war plans officer. His efforts won the approval of Marshall, who sent him to Britain in the summer of 1942 in command of US forces. In London, he passed the most demanding test of all by impressing Winston Churchill, who approved of his appointment to command Allied forces for Operation Torch, the joint Anglo-American invasion of French North Africa in November. In 1942 and 1943 Eisenhower commanded Allied forces in the Mediterranean. The campaigns in Tunisia, Sicily and Italy were all testing but valuable experiences that led to his appointment in December 1943 as supreme Allied commander for the invasion of northwest Europe in 1944.

Operation Overlord

In 1944 Eisenhower came of age as the commander responsible for the planning and execution of the largest amphibious invasion ever mounted, Operation Overlord, the invasion of Normandy, originally scheduled for 5 June. The Allies had massed nearly 6,000 ships and 150,000 men when a full-blown gale not only rendered any hope of launching the invasion on the morning of 5 June unthinkable, but also threatened to wreck the entire invasion timetable.

Eisenhower was compelled to order a postponement while the armada trod water, its thousands of participants virtual prisoners in their encampments and aboard naval vessels; final briefings were postponed and sealed instructions revealing their target remained unopened. A mood of pessimism prevailed among many senior Allied commanders that in spite of the extensive preparations and training the invasion might still fail on the beaches of Normandy. The atmosphere was not lightened by updates from Allied intelligence that the German commander, Field Marshal Erwin Rommel, had strengthened the Normandy front by several new divisions, with more possibly on the way.

At the late evening briefing on 4 June the assembled generals, admirals and air marshals could distinctly hear the sounds of pounding rain and the wind howling outside. Nevertheless, the senior Allied meteorologist, RAF Group Captain J. M. Stagg, reported to the tense commanders that there was a glimmer of hope for 6 June. While the weather would remain poor, visibility would improve and the winds decrease, though barely enough to risk launching the invasion. This was arguably the most important weather prediction in history: a mistaken forecast for D-Day could turn the entire tide of the war in Europe against the Allies. Eisenhower swiftly learned that time had run out. He had to make a crucial decision for or against, then and there. He was obliged to weigh not only the decision itself but its longer-term impact. There was utter silence in the room except for the sound of the wind and rain as Eisenhower pondered whether or not to permit the invasion to go forward. Any delay beyond 6 June meant postponement of Overlord until the next full moon in July, further risking its compromise by the Germans.

In preparation for this eventuality, Eisenhower and his weather team had practised such a scenario for some weeks. Finally, he announced his decision: 'OK,' he said. 'We'll go.' With that simple declaration, Eisenhower had made his historic decision. Once unleashed, the Allies were irrevocably committed and it took considerable courage to set into motion the operation that would decide the victor and the vanquished of the war. Had any been present, Eisenhower's critics, who have painted him in unflattering terms as a chairman of the board, beholden to many and in command of none, would have witnessed his finest hour.

Rommel's naval deputy, Vice Admiral Friedrich Ruge, later marvelled that Eisenhower made such an important decision without recourse to higher authority, noting that no one in the German chain of command would have dared. It was, Ruge believed, 'one of the truly great decisions in military history'.

On 6 June (D-Day) Allied forces crossed the English Channel to seize beachheads on the Normandy coast (see also pp. 229–30). Although the fighting in the close hedgerow country (*bocage*) was difficult, the Allies succeeded in breaking out in August. Paris was liberated on 25 August. Most of France, Belgium, Luxembourg and the southern Netherlands quickly fell into Allied hands, but the advance ground to a virtual halt in mid September with the failure of Montgomery's attempt to seize bridgeheads at Arnhem on the lower Rhine (Operation Market Garden; see pp. 346–7).

The Battle of the Bulge

On 16 December 1944, with the Allies stalled on the border of Germany, Hitler, in an all-out gamble to compel the Allies to sue for peace, launched the largest German counter-offensive of the war in the West, with three armies, more than a quarter of a million troops. Its objective was to split the Allied armies by means of a surprise blitzkrieg thrust through the rugged, heavily forested but lightly defended

Ardennes, a planned repeat of what they had done twice previously – in August 1914 and May 1940.

American units were caught flat-footed and fought desperate battles to stem the German advance. The US 101st Airborne Division was surrounded at Bastogne, its commander issuing a one-word refusal to surrender that has become a symbol of defiance: 'Nuts!' As the German armies drove deeper into the Ardennes in an attempt quickly to secure vital bridgeheads west of the River Meuse, the line defining the Allied front on the map took on the appearance of a large protrusion or 'bulge', the name by which the battle would forever be known.

Eisenhower made another courageous and controversial decision, this one to appoint British field marshal Bernard Montgomery to take command of the northern half of the 'Bulge' and restore order over what was a predominantly American sector. The decision proved far-reaching and at the height of the battle Eisenhower was promoted to the five-star rank of General of the Army. Although the Battle of the Bulge (16 December 1944 – 25 January 1945) earned the dubious distinction of being the costliest engagement ever fought by the US Army, which suffered over 100,000 casualties, the gallantry of American troops fighting in the frozen Ardennes proved fatal to Hitler's ambition somehow to snatch, if not victory, at least a draw with the Allies. *Time* magazine named Eisenhower its 'man of the year' for 1944.

At 2.41 a.m., 7 May 1945, Germany surrendered unconditionally at Supreme Headquarters Allied Expeditionary Force (SHAEF) in Reims, France. As Colonel General Alfred Jodl signed the documents of surrender for Germany, conspicuously missing was the man who had orchestrated the events leading to this historic moment. The death of soldiers in the cause of peace was close to Dwight Eisenhower's heart and soul. He was not unique as a professional soldier who hated war; but few hated their enemy with greater passion than did Eisenhower, who regarded his adversaries with nothing short of loathing. At Reims, he declined to attend the surrender ceremony and instead delegated to his chief of staff, Lieutenant General Walter Bedell Smith, the task of signing the surrender documents.

As a soldier, Eisenhower understood that it was not his place to announce the end of the war in Europe, but a function of the heads of state who would make the formal announcement the following day. It was typical of Eisenhower that he would not take credit for the Allied victory. Instead his message to his bosses – the combined chiefs of staff – was utterly devoid of self-congratulation, and as unpretentious as the man himself. Only a single sentence long, it read simply: 'The mission of this Allied force was fulfilled at 0241 hours local time, May 7, 1945. [signed] Eisenhower.'

Only after the Germans had departed did Eisenhower finally unbend and relax. As a horde of photographers were admitted to his office and scrambled to record the scene, his famous grin reappeared and he signalled a 'V' for Victory by holding aloft the two gold pens with which the German surrender documents had been signed.

For all his military experience, Ike ultimately came to detest war and everything it stood for. He once said, 'I hate war as only a soldier who has lived it can, only as one who has seen its brutality, its futility, its stupidity.' That Eisenhower well understood that sometimes war is inevitable is beyond question. Yet so profound was his experience that – as president of the United States – he was moved to remark in 1953: 'Every gun that is made, every warship launched, every rocket fired signifies, in the final sense, a theft from those who hunger and are not fed, those who are cold and are not clothed.'

The post-war years and the Eisenhower presidency

Eisenhower emerged from the war a national hero. He served as US Army chief of staff from 1945 to 1948, when he became the president of Columbia University. His memoir of the war, *Crusade in Europe*, became an international bestseller and greatly enhanced his public image. In December 1950 he was recalled to active duty and appointed the first supreme commander of the North Atlantic Treaty

Organization (NATO), where he served until May 1952 when his military service ended. His great national popularity led to his drafting as the Republican party candidate for president and his election in November 1952, and his re-election by a landslide margin in 1956.

Dwight Eisenhower's trademark grin was perhaps the best known of any public figure in American history. *Time* magazine said of him in 1952: 'They saw Ike, and they liked what they saw ... They liked him in a way they could scarcely explain. They liked Ike because, when they saw him and heard him talk, he made them proud of themselves and all the half-forgotten best that was in them and in the nation.'

His two terms as one of the most popular presidents in American history were marked by passage of the first civil rights acts since the end of the Civil War, intervention in Arkansas to enforce desegregation of the Little Rock public schools, and the promulgation of the Eisenhower Doctrine of American intervention against any nation that used armed force to promote international communism. One of Eisenhower's enduring achievements as president was the creation of the Interstate Highway System in 1956. Its enabling legislation came about as a result of his experience during the Transcontinental Motor Convoy of 1919.

Eisenhower's legacy

Dwight Eisenhower may not have fitted the mould of the warrior hero or of a battlefield general in the tradition of Robert E. Lee, 'Stonewall' Jackson or George S. Patton, yet he was every inch a soldier. His legacy is based on effectively shaping an alliance of Britain and America, two prickly, independent-minded countries with fundamentally disparate philosophies of waging war.

History has unfairly stereotyped Eisenhower as a lightweight and a bumbling orator, famous for mangling his syntax, who spent his lifetime concealing his intellect and intelligence behind the image of the 'country bumpkin from Kansas'. But the famous grin masked a man of great intelligence, whose hidden-hand leadership was that of a master craftsman and a thoughtful, caring human being. Eisenhower did far more than make Americans proud. As a soldier and a commander, his place in history was indisputably earned as the supreme commander who guided the Allies to victory in the Mediterranean and in Europe during the most destructive war in history.

Whether in war or peace, Eisenhower always insisted there was no such thing as an indispensable man. Twenty years after the war he was aboard the liner *Queen Elizabeth* for a nostalgic return to Normandy, the scene of his greatest triumph. One night over dinner he said he had read a poem that summed up his attitude about indispensability. It ended this way:

> The moral of this quaint example
> Is to do just the best that you can.
> Be proud of yourself, but remember,
> There is no indispensable man.

Perhaps. But it can be safely argued that in the Second World War, Dwight David Eisenhower was indispensable to the victory of the Allies.

GEORGE S. PATTON

1885–1945

TREVOR ROYLE

GEORGE SMITH PATTON was the outstanding exponent of armoured warfare produced by the Allies in the Second World War. During the fighting in Europe in 1944 and 1945 he demonstrated a mastery of handling large armoured formations, deploying them in mobile, high-speed operations, and thereby winning the admiration of German opponents such as General Günther Blumentritt, who called him 'the most aggressive panzer-general of the Allies'. An inspired leader who possessed physical and mental energy in abundance, Patton was blessed with courage and boundless self-confidence, virtues that made him a spirited and highly effective military commander who always encouraged his soldiers to embrace the will to win. But for a moment of madness in Sicily in August 1943, when he struck two private soldiers suffering from shell shock and was duly reprimanded, he would have commanded the US ground forces following the D-Day landings instead of his great rival, General Omar N. Bradley.

Very much a product of his times, Patton was imbued with his country's military doctrine of speed, manoeuvre and surprise, followed by the deployment of overwhelming force to crush the opposition. It was simple and effective, and the philosophy is best summed up by Patton in his own words: 'The only way you can win a war is to attack and keep on attacking, and after you've done that, keep attacking some more.' He was also a larger-than-life character whose flamboyance made him an unmistakable figure in an increasingly dull and monotonous age. At a time when many commanders dressed in the same style as their soldiers and were prone to disguise trappings of rank, Patton sported a smart uniform complete with stars, riding boots and ivory-handled revolvers, and his flashy motorcades always made sure that he was the centre of attention.

Early training

Patton was born on 11 November 1885 and received his military training at the Virginia Military Institute in Lexington and at West Point. In 1909 he was commissioned in the 15th Cavalry Regiment, and three years later represented the USA in the modern pentathlon event at the

Olympic Games in Stockholm. His first chance to test himself as a soldier came in a frontier war that broke out along the border with Mexico in the summer of 1916. It was a minor conflagration involving rival gangs of Mexican bandits and there were no set-piece battles, but the experience was not lost on Patton. Not only did it provide him with his first opportunity of coming under fire, but having been appointed to the staff of General John J. Pershing, he also learned about the importance of good staff work. In most respects the Mexican campaign was a series of low-intensity operations against a mobile enemy who used the vast reaches of the border area to good advantage by mounting attacks on US positions and then disappearing into thin air. Matters came to a head in March 1916 when Pancho Villa, a Robin Hood-like desperado, attacked the town of Columbus in New Mexico and killed eighteen Americans, including six soldiers.

To prevent further incursions a punitive expedition consisting of 10,000 soldiers was mounted under Pershing's command. Although it failed to crush Villa, it marked the beginning of a new phase in US military operations – the force was motorized and accompanied by eight army biplanes for reconnaissance. The expedition also added lustre to Patton's name. In a daring raid at San Miguelito in May 1916 Patton's patrol ambushed Villa's principal lieutenant, Julio Cárdenas, who was killed along with two other Mexicans in a shoot-out that could have come straight out of a Western movie.

First World War service

The US decision to declare war on Germany in the spring of 1917 gave Patton his first opportunity to experience modern industrialized warfare. Initially Patton served on the staff of the American Expeditionary Force (AEF), but on 10 November he was ordered to take command of a training school for tank crews as part of the AEF's initiative to create a tank corps of 200 heavy and 1,000 light tanks.

The main drawback was that the AEF was reliant on the French for its supplies of Renault FT-17 tanks, and it was not until the spring of 1918 that they began arriving. In April Patton was promoted to lieutenant colonel and given command of the US army's first two light-tank battalions, which he quickly welded into a coherent and well-disciplined force. Because the tank crews would be fighting on a battlefield in which split-second decisions would have to be taken and they would be operating vehicles without any means of communication, Patton insisted that they maintain high standards of discipline. His message was simple: if his tank commanders assimilated close-order drill and if it became second nature to them they would produce 'instant, cheerful, unhesitating obedience' in the chaos of the battlefield.

In the late summer of 1918 Patton's tank force, the 1st US Tank Brigade (344th and 345th Tank Battalions), saw its first action in the Battle for the St Mihiel salient, a triangular bulge in the German lines that reached as far as the River Meuse south of Verdun. The feature had been in German hands since the first months of the war, producing a threat to the French forces in Champagne, and it was agreed by the French that it represented a good objective for an independent American action on 12 September. Patton's tank brigade was ordered to support the attack on the southern salient by two American infantry divisions, the 1st and the 42nd. As the tanks did not start arriving until 24 August Patton had little time to arrange the supporting fuel and ammunition dumps, and it was not until two hours before the attack began that he was able to get his last tanks into position. He had also reconnoitred the ground over which his men would fight, a requirement that he would impose on all of his commanders later in his career.

Compared to many other battles fought on the Western Front, the offensive at the St Mihiel salient was a minor affair. It only lasted two days, and the US success was as much due to the German decision to withdraw as to the skill of the AEF, but the attack boosted the morale of those who took part in it. The US Tank Brigade also came of age. The two battalions led the attack; of the 174 tanks that took part in the fighting only three received hits and, remarkably, only forty-three broke down. Some got too far ahead of the infantry while others foundered in the heavy-going ground, but the tanks showed that they were capable of winning and holding terrain. In one attack, a section of tanks broke the German line, thereby proving Patton's point that they could be used as a mobile penetration force capable of smashing into the enemy's rear.

Armoured warfare

Although the AEF came back to a hero's welcome, the huge expeditionary force was quickly disbanded, and by the beginning of 1920 the army was once more small in number, having reduced its size to just 130,000 soldiers. Patton stayed with the Tank Corps until the autumn of 1920, when he returned to the cavalry, serving twice in Hawaii and in Washington between 1928 and 1934, when he worked on the staff of the chief of cavalry. During the 1920s and 1930s he wrote and published extensively, producing a wide variety of papers on mechanized warfare, manoeuvre in battle, sport as training for war and, above all, the need for personal leadership. In 1923 he completed the field officers course at Fort Riley, Kansas, and the following year he graduated from the Command and General Staff College at Fort Leavenworth, also in Kansas. This was followed in the winter of 1931/2 by a spell at the Army War College at Fort McNair, Washington, DC.

By 1938 Patton had been restored to his wartime rank of full colonel; with it came command, first of the 5th Cavalry Regiment and then the 3rd Cavalry Regiment. By then, too, another war seemed probable, and for the next three years the US army underwent a period of rapid modernization. New armoured forces were raised and Patton was given command of the 2nd Armored Division, which quickly became a byword for its skills and efficiency. During a series of war games played out in Tennessee, Louisiana and South Carolina in 1941 his division excelled, and shortly after the USA declared war on Japan and Germany at the end of the year Patton was given command of I Armored Corps.

During this period of intensive training a pattern had begun to emerge in Patton's style of leadership. He insisted on speed and aggression in making attacks, and tried to gain the upper hand by relying on surprise, using reconnaissance to good advantage – he was a stickler for gathering intelligence about his opponents' movements. He also inculcated discipline in his commanders so that their standards would be passed on to the men, and made sure that he was known to all his soldiers, devoting considerable time to addressing them in person. On the technical side he was an innovator, flying a light aircraft to tour the operational areas and using wireless to keep in touch with forward positions. In short, by the time the USA went to war, Patton had proved himself to be an inspirational commander capable of leading large numbers of men in battle.

North Africa and Sicily

Following the US declaration of war it was agreed that US forces should lend their weight to the Allied campaign in North Africa. Known as Operation Torch, its object was to gain complete control of North Africa from the Atlantic to the Red Sea, starting with landings in Algeria and French Morocco before rolling up the Axis forces in Libya, where the British Eighth Army was engaging the Axis forces. Patton was given command of the task force that landed at Casablanca unopposed on 8 November 1942. Thereafter matters did not run so smoothly, and just two months later Major General Lloyd Fredendall's US II Corps suffered ignominious defeats at Sidi-Bou-Zid and the Kasserine Pass in February 1943 (see pp. 218–19). As a result Fredendall was sacked and replaced in March by Patton, who quickly set about the essential task of restoring discipline and rebuilding morale.

Within ten days of taking command Patton was leading his men in the Allies' first joint offensive to crush Axis resistance in North Africa. The plan was to use the British

Eighth Army to attack Tunis through the Mareth Line from the south while a New Zealand army corps made a diversionary attack inland to outflank the Axis defensive positions. If the assault succeeded, the Italians would be driven north along the coastal plain towards Tunis, where they and the German forces would be crushed by the Allies. Patton's role in the operation was to attack enemy positions along the eastern flank – running through the hills known as the Eastern Dorsale – and to capture the hill towns of Gafsa, Sened and Maknassy, together with the associated air strips that would otherwise threaten the Eighth Army as it drove towards Tunis. All the objectives were taken with an élan that would have been impossible in the immediate aftermath of the débâcle at the Kasserine Pass.

On 13 May 1943 Tunis fell, and the war in North Africa came to an end with a resounding victory that did much to restore Allied self-respect. The next stage involved the capture of Sicily as a precursor to the invasion of Italy. The Sicilian operation called for a seaborne assault by the British Eighth Army between Syracuse and the Pachino peninsula on the island's southeastern coast on 10 July, while the US I and II Armoured Corps under Patton's command would land on a 40-mile front along the southern coast between Gela and Scoglitti and Licata on the left flank. British and US forces also staged an airborne assault to capture key points. The general plan was that the British would push north on the right flank while Patton's US forces shielded the left.

This was a difficult time for Patton, who believed that US forces were being given a subsidiary role – a symptom of the friction then prevalent in the Allied command structure. Patton's solution was to give a demonstration of his army's capability by rushing northwestwards to take possession of the capital, Palermo. In itself the city was not a vital strategic objective – the main aim was to take Messina in the northeast corner – but Patton knew that a successful operation would increase the status of the US army. On 21 July Palermo duly fell, and this allowed Patton to use it as a springboard for attacking Messina, which was entered by triumphant US forces on the evening of 16 August.

France and northwest Europe

Patton's successes in North Africa and Sicily should have been followed by promotion to command the US forces for the invasion of Europe (see pp. 229–30 and 272–5), but following the incidents in which he slapped two privates he was sidelined and given command of the US Third Army. It was not until the end of July 1944, almost two months after the D-Day landings, that Patton joined the battle in Normandy – but he immediately made his mark.

Following the successful Allied landings in Normandy in June 1944, the next phase of the operation involved the Allies fighting to consolidate their beachhead before attempting a break-out, while the Germans made every effort to drive them back into the sea. This gave Patton, who arrived in Normandy in late July, his chance to exploit the situation by leading a rapid armoured assault to move US forces into open country before the Germans had the chance to regroup. It was to be the only example of the use of blitzkrieg by the Allies during the Second World War – the rapid and ruthless penetration of the enemy's lines using over-whelming force to encircle and destroy the opposition. Air support was integral to this style of warfare, with strike aircraft acting as airborne artillery in advance of the main tank assault.

The Normandy stalemate was broken by Operation Cobra, mounted on 25 July by the US First Army, which pushed as far south as Avranches and the pivotal neighbouring town of Pontaubault. Suddenly the possibility opened of invading Brittany in the west and racing eastwards towards Le Mans and the River Seine. The task was given to seven divisions of Patton's US Third Army, which moved with exemplary speed into Brittany, frequently running ahead of their lines of communication as they sped into the open countryside. Bottlenecks and traffic jams were overcome by the simple expedient of dispatching staff officers to forward positions with instructions to get the units through, regardless of their sequence in the battle-plans.

Within three days Patton's divisions were through the Avranches–Pontaubault gap; it was not a manoeuvre that would have been recognized at staff college, but it worked.

While this breakthrough was gratifying it was obvious that Brittany had become a backwater and that there was better employment for Patton's forces. They had shown what could be accomplished by a mobile army using speed and aggression backed up by air power; now was the time to deploy those assets eastward against the bulk of the German forces guarding the Paris–Orléans gap. It was there, argued Patton, that 'the decisive battle of the European war would obviously be fought'. Backed by fighter-bombers of the US XIX Tactical Air Command, the US Third Army pushed through the southern Normandy countryside, advancing rapidly without worrying about the need to protect their flanks.

Once engaged in the break-out, Patton's armoured divisions had shown that they had taken their general's philosophy to heart. What began at Avranches continued in the race to the Seine, and the operation was to demonstrate everything that was good about Patton's leadership in the pursuit phase of a battle. With the fall of Dreux, Chartres and Orléans, all the objectives had been achieved by mid August, leaving Patton to exult that the operation was 'probably the fastest and biggest pursuit in history'.

By 30 August Patton had moved the Third Army across

the River Marne and was within reach of Metz when their petrol reserves began to run dry. Only the lack of regular supplies – the logistic tail stretched over 400 miles back to Cherbourg – prevented Patton from pressing home his advantage, and for a time the Allied advance stalled.

It was at this stage of the battle, when the Allies were still confident that the end of the war was in sight, that the Germans decided to counter-attack in the Ardennes. The plan was the brain-child of Adolf Hitler, who reasoned that the winter weather – 'night, fog and snow' – would give the Germans the opportunity to hit back at the Allies through the dense Ardennes forest, with its narrow steep-sided valleys, and then turn rapidly north to recapture Brussels and Antwerp. It did not turn out that way, but the Battle of the Bulge, as it came to be known, almost allowed the Germans to achieve their aims by creating a huge salient or 'bulge' in the Allied lines. During the battle, in one of the best-executed manoeuvres of the war, Patton realigned his Third Army, turning its three divisions to attack northwards out of Luxembourg to dent the German assault. Owing to Patton's foresight – he had already anticipated the move and had plans in place to execute it – the entire operation was completed within four days. Once the bulge had been blocked German resistance in the Ardennes came to an end.

Ahead lay the drive into Germany and the crossing

of the River Rhine, as the Allies pushed into the enemy heartlands. Patton's Third Army raced into southern Bavaria and ended the war close to the border with Czechoslovakia, but its commander did not live long to relish the triumph or to take his military career into the post-war world. It is one of the many ironies in his life that Patton did not die, as he wished, 'with the last bullet in the last battle of the war', but as the result of a needless car crash near Mannheim in Germany. He died on 21 December 1945 in a hospital in Heidelberg, and was buried at the US military cemetery in Luxembourg, his grave lying alongside other Third Army soldiers who had died under his command a year earlier.

TOMOYUKI YAMASHITA

1885–1946

ALAN WARREN

LIEUTENANT GENERAL TOMOYUKI YAMASHITA's seventy-day campaign to capture Singapore on 15 February 1942 was one of the most spectacular in modern history, earning him the nickname 'the Tiger of Malaya'. At Singapore, the British Empire's illusion of permanence and strength was brought crashing down in a matter of weeks. A rational strategic thinker, Yamashita was not a man obsessed with spiritual values at the expense of practical considerations. With a large, shaven head and broad neck, he was tall by Japanese standards, a dignified man noted for his stoicism and lack of public emotion. Behind the mask, however, he was energetic, ambitious and possessed of a high level of imagination. A senior staff officer shrewdly described Yamashita as a 'clear-headed type of politician'.

War in Asia had broken out in July 1937 as a consequence of Japanese encroachments on northern China. By 1939 the conflict between Japan and the Chinese Nationalists had reached stalemate. Yet in the wake of Germany's stunning triumph over France, in September 1940 the Japanese government opportunistically joined the Axis and became an ally of Nazi Germany. The situation was transformed again when the Germans invaded the Soviet Union in June 1941. With the Russians, Japan's traditional enemy, fully occupied, Japan's imperialist leaders saw chances for conquest at the expense of poorly defended European colonies in Southeast Asia. When the Japanese were drawing up plans to overrun the region, the best available army formation – the Twenty-fifth Army – was set aside to attempt the capture of Singapore, the most important Allied base in the region.

Early career

The general appointed to command the vital Twenty-fifth Army was Lieutenant General Tomoyuki Yamashita. The second son of a doctor, Yamashita had been born in November 1885 in an isolated mountain village on the Japanese home island of Shikoku. He attended the Central Military Academy in Tokyo, and was commissioned at the end of 1905.

In 1918, after graduation from the War College,

Yamashita arrived at the Japanese embassy in Switzerland as assistant military attaché. Captain Hideki Tojo, Japan's future wartime prime minister, was also posted to the Berne embassy at that time. The two men struck up a friendship and, after the end of the First World War, toured Germany and the silent battlefields of the Western Front. Yamashita returned to Tokyo and the War Ministry, but he later spent a further three years in Europe as military attaché in Vienna, leaving his estranged wife behind in Japan. While in Vienna, Yamashita had an extended and passionate affair with a German woman; he later said that his Vienna posting was the best period of his life.

By 1936 Yamashita was a major-general and chief of military affairs at imperial Headquarters in Tokyo. An attempted military coup that year badly damaged his political position within the army, and drew the ire of the emperor. Henceforward Yamashita was keenly aware of that disapproval and the consequent need to demonstrate his loyalty publicly. A spell in Korea as a brigade commander was followed by service as a divisional commander in China after war broke out in 1937. Yamashita, on active service for the first time, earned a fine reputation for bravery and leadership.

After Japan had joined the Axis, Yamashita led a military mission to Europe to investigate Germany's crushing victory over France. He met Hitler in Berlin and said privately that he was 'an unimpressive little man'. Yamashita was exposed to German concepts of armoured warfare that emphasized speed and mobility. After returning to Japan, he was posted to Manchuria as part of the build-up in that region in preparation for possible war with the Soviets. Early in November 1941 he was recalled to take command of the Twenty-fifth Army for the invasion of Southeast Asia.

The invasion of Malaya

In 1941 Singapore, the greatest port between India and China, was part of the British colony of Malaya. The British had built a naval base at this 'Gibraltar of the East', and emplaced large-calibre naval guns on the south coast of the island to deter an attack from the sea. There was, however, no British fleet there prior to the Second World War. Owing to the demands of the war in Europe, only two British capital ships – *Prince of Wales* and *Repulse* – could be spared for the Far East.

The British Empire's army garrison in Malaya had been strongly reinforced during 1941 to over 90,000 men, yet there was no British tank force in the colony. The army commander in Malaya was Lieutenant General A. E. Percival, a mild-mannered man who had spent much of the previous decade as a desk-bound staff officer.

At Southern Army's Saigon headquarters, General

Count Hisaichi Terauchi gave the Twenty-fifth Army his best divisions. Japanese officers of the Taiwan Army Research Section had been working on preparations for the campaign in Southeast Asia for months. Amphibious landings had been practised in southern China and on Hainan Island, and exercises undertaken using bicycles to convey combat units long distances.

General Yamashita left Saigon for Hainan Island on 25 November 1941, where a part of the Twenty-fifth Army was assembling at the port of Samah. On 4 December, as his forces set out from Hainan, Yamashita wrote a poem:

> On the day the sun shines with the moon
> Our arrow leaves the bow
> It carries my spirit towards the enemy
> With me are a hundred million souls
> My people from the East
> On this day when the moon
> And the sun both shine.

The finalized Twenty-fifth Army plan was to land the 5th Division at Singora and Patani in southern Thailand, whilst a regiment of the 18th Division landed at Kota Bharu in northeast Malaya to seize nearby aerodromes. The Imperial Guards Division was to take part in the invasion of Thailand, and then move down into Malaya by road. A brigade-strength tank unit was available to support the Twenty-fifth Army's infantry. Japanese troops were experienced, obedient and possessed an indestructible morale that drew inspiration from *bushido*, the Spartan creed of the Samurai.

On 7 December 1941 Japan's daring plan to attack the United States fleet at Pearl Harbor came to fruition, together with the invasion of Thailand and Malaya. Yamashita and his headquarters were aboard the transport ship *Ryujo Maru*, and he intended to go ashore at Singora, close behind the first wave of invaders. The convoys were only briefly sighted as they closed on their destinations. The landings in southern Thailand were almost unopposed, and the Thai government soon capitulated. The landing on the north-east coast of Malaya at Kota Bharu was timed to start at the same moment as the raid on Pearl Harbor, though by accident it actually began an hour or two early. At the outset of the campaign, the under-strength RAF was quickly shot out of the skies of northern Malaya, while *Prince of Wales* and *Repulse*, attempting a delayed foray in the South China Sea, were swiftly sunk by modern Japanese torpedo-bombers.

On land, the well-balanced Japanese expeditionary force advanced south from Thailand. At Jitra, in northwest Malaya, the Anglo-Indian defenders were swept aside, and Yamashita sensed that he already had the British on the run. Japanese troops used bicycles to keep moving forward along good local roads. General Percival had retained half of his force at Singapore and southern Malaya as he feared a direct amphibious attack on those places. Penang was abandoned, and the Japanese used small boats to 'hook' troops down the west coast of Malaya. The dazed 11th Indian Division was shattered by Japanese tanks at the Slim River on 7 January 1942, and Kuala Lumpur, the capital of the Federated Malay States, fell to the invaders a few days later. In southern Malaya the enterprising Japanese broke through a weakly held flank and the British retreated to Singapore Island. The Japanese had advanced over 500 miles in less than eight weeks, and had suffered fewer than 5,000 casualties.

Towards the close of January, several Japanese air force units were transferred from Malaya to support operations in the Dutch East Indies. An angry Yamashita had responded: 'All right, in that case we shall not rely upon the cooperation of the air force. The army will now capture Singapore single-handed.' British reinforcement convoys continued to sail into Singapore at the eleventh hour, bringing the garrison to over 100,000 men. A vast pillar of dark smoke billowed skywards from the burning fuel dumps of the evacuated naval base. The Battle of Singapore was about to begin.

General Yamashita resolved to attack Singapore in

February 1942 after only a week's preparation. The Twenty-fifth Army's advanced headquarters was sited in the Sultan of Johore's Green Palace, the Istana Hijau, on the bank of the Malayan mainland facing Singapore. Yamashita sent a message to his divisional commanders telling them that he would directly observe their efforts from his new command post. He assured his adjutant: 'The enemy won't fire on this place. They would never dream I would come so near in such a prominent position.' Yamashita's leadership was, as always, energetically driving his army forward.

The strait separating Singapore from the Malayan mainland is narrow, only half a mile to a couple of miles across. Yet, rather than concentrate a large part of his force opposite where the strait was narrowest, the British commander, General Percival, spread his garrison right around Singapore's coast. By use of various decoys Percival's attention was diverted to the northeast coast of Singapore, and on the evening of 8 February 1942 Yamashita threw just about his entire force against the northwest coast in 300 small boats. The force deployed by Percival along that stretch of coast consisted of just one Australian brigade. By morning the thin line of defenders had collapsed and retreated several miles to the rear in disorder. The following night the Imperial Guards attacked between the Kranji River and the causeway linking Singapore Island to the mainland. The assault of the Imperial Guards came close to failure, but

Yamashita refused to be panicked and the attack was pushed onward successfully.

Percival was still worrying about fresh landings on other parts of the Singaporean coast, unaware that almost the entire Japanese force was already ashore. Yamashita's aggressive handling of his forces made them seem numerically stronger than was in fact the case. On the night of 10/11 February the Bukit Timah heights, in the centre of Singapore Island, were seized in a bayonet assault. After the Japanese managed to land tanks they consolidated their hold on the central part of Singapore Island, including the reservoir catchment area.

From 12 February constant Japanese attacks slowly contracted the British perimeter. As Japanese ammunition ran short, Yamashita visited his units to apologize for the shortage and to urge continued advance relying on the bayonet. Yamashita told his staff: 'The enemy is also going through a hard time. If we halt now, we will lose the initiative over the battle and the enemy will discover our shortage of supplies and counter-attack us!' When Percival finally agreed to negotiate a surrender on 15 February, the beaten garrison had been backed into a narrow perimeter surrounding the town and waterfront.

The capitulation of Singapore

Yamashita insisted on a face-to-face meeting with Percival

to conclude the week-long battle for Singapore, and on 15 February 1942 the British commander led a small party bearing the Union Jack and a large white flag to parley with the Japanese. The conference, which began at 5.15 p.m., took place at the Ford Motor Factory near Bukit Timah village. Percival looked haggard; Yamashita presented a bullet-headed, stocky figure in his khaki field uniform. When discussions bogged down, Yamashita impatiently banged the table with his fist: 'The time for the night attack is drawing near. Is the British army going to surrender or not? Answer YES or NO?' (He used the English words.) Japanese newspapers reported exaggerated accounts of Yamashita bullying Percival into surrender with his bulk and glaring eyes.

Yamashita wrote that Percival 'was good on paper but timid and hesitant in making command decisions', and said of the final stage of the campaign: 'I knew if I had to fight long for Singapore I would be beaten. That is why the surrender had to be at once. I was very frightened all the time that the British would discover our numerical weakness and lack of supplies and force me into disastrous street fighting.' Throughout the campaign Percival had failed to concentrate his forces, whereas the Twenty-fifth Army had massed force on a narrow front, and made full use of surprise and manoeuvre to punch through its enemy's line. Yamashita's plan for the assault on Singapore Island, whereby the bulk of the Twenty-fifth Army was thrown against a small part of the defending force, repeated the formula that had worked so well on the mainland.

A ceasefire came into effect at 8.30 p.m. on 15 February. The following day a vast army of British, Indian and Australian servicemen went into grim and often fatal captivity. For the British, the shame of the manner of Singapore's fall demonstrated, in no uncertain terms, that the days of empire and world power were drawing to a close. At dawn on 16 February, the day after the capitulation, Yamashita rose from his bed and went outdoors. He stood at the edge of a wood and bowed to pray, facing in the direction of the emperor's palace in Tokyo. The Japanese High Command had allotted a hundred days to take Singapore. The task had been successfully completed in seventy.

One of the more immediate consequences of the fall of Singapore was the summary execution of thousands of Chinese by Japanese forces in the first weeks after the surrender. The Singapore garrison commander, Major-General Kawamura, subsequently claimed Yamashita gave the order on 18 February for mopping-up operations against 'hostile' Chinese, or what became known as the *Sook Ching*, 'purification by elimination'. Senior Japanese officers later conceded that 5,000 Chinese had been executed by military police (*Kempaitai*) and detachments from all three of the 25th Army's divisions. Leaders of Singapore's Chinese community have estimated that over 30,000 people were killed.

The battle for the Philippines, 1944–5

Yamashita did not entirely approve of his popular sobriquet, 'the Tiger of Malaya', and once told a German military attaché: 'The tiger attacks its prey in stealth but I attack the enemy in a fair play.' Surprisingly, Yamashita did not play a prominent part in the war in the period following his capture of Singapore. In July 1942 he was sent from Singapore to command one of the two army groups stationed in Manchuria along the frontier with the Soviet Union. This was a command of importance, but within the army it was felt that Yamashita had been kept away from Japan by Prime Minister Tojo for political reasons. Yamashita had not been permitted to stop off in Japan en route to Manchuria for the audience with the emperor that a victorious military commander had reason to expect.

Only after Tojo's resignation as prime minister in July 1944 was Yamashita recalled from Manchuria. He finally had an audience with Emperor Hirohito on 30 September 1944 and was told that 'the fate of the empire rests upon your shoulders'. Yamashita was swiftly dispatched to lead Japanese forces in the Philippines. He assumed command of over 400,000 men on 9 October 1944, just as the Americans were poised to attack. The situation for Japan had changed dramatically for the worse since the fall of Singapore.

The first American landings were made on the island of Leyte on 20 October 1944 amid a great series of naval battles. By the end of 1944 Japanese forces on Leyte had been destroyed by the massive firepower the Americans were able to deploy. Luzon, the main Philippine island, was invaded on 8 January 1945. Yamashita had no intention of defending Manila, the capital of the Philippines, and ordered his local army commander to conduct demolitions ahead of the advancing Americans and then evacuate the city. At Manila, however, Rear Admiral Sanji Iwabuchi decided that it was his duty to deny Manila's harbour installations for as long as possible. In the battle that followed (February–March 1945), the Japanese force was annihilated, but parts of the city were demolished, an estimated 100,000 civilians died, and there was widespread murder, rape and torture of civilians by Japanese troops within the confined space of the shrinking perimeter.

In north Luzon, Yamashita's main force tenaciously held its ground until crippling losses drove it deep into the mountainous interior. After Japan's final capitulation, a thin Yamashita gave himself up to the Americans on 2 September. He did not commit hara-kiri and was reported to have said: 'If I kill myself someone else will have to take the blame.' There is no doubt that Yamashita's defence of Luzon displayed generalship of the highest standard.

Trial and execution

Yamashita was put on trial as a war criminal by the Americans. The trial in Manila began on 29 October 1945. The main charges against the general related to his responsibility for the massacres that had undoubtedly taken place in Manila. Senior Japanese officers testified that Yamashita had ordered that Manila was not to be defended. Nonetheless, the general's death sentence was pronounced on 7 December, the anniversary of the bombing of Pearl Harbor. Yamashita told an interviewer:

> My command was as big as MacArthur's or Lord Louis Mountbatten's. How could I tell if some of my soldiers misbehaved themselves? It was impossible for any man in my position to control every action of his subordinate commanders, let alone the deeds of individual soldiers ... What I am really being charged with is losing the war. It could have happened to General MacArthur, you know.

Yamashita was hanged on 23 February 1946. The dubious legal circumstances of his conviction have garnered for him a degree of sympathy, and were seen in some quarters as an official lynching. Yet if the general had not been tried as a war criminal in Manila, he would very likely have been tried and executed by the British in Singapore for the organized massacre of the Chinese carried out by Japanese troops after their capture of the island.

DOUGLAS MACARTHUR

1880–1964

CARLO D'ESTE

ONE OF THE GREATEST SOLDIERS ever to wear the US Army uniform and one of the most enigmatic, MacArthur's flamboyance and brilliance were matched by a towering ego and a penchant for the dramatic. His star-studded career spanned half a century: first in his class at West Point, most decorated combat commander of 1917–18, superintendent and reformer of West Point, chief of staff of the inter-war army, military proconsul both in the Philippines and in Japan, Medal of Honor winner, commander of Allied forces in the South Pacific, commander of UN forces during the first critical months of the Korean War, and one of only five men to be accorded the five-star rank of General of the Army.

Douglas MacArthur was born in Little Rock, Arkansas, in 1880, the son of a career army officer, Lieutenant General

Arthur MacArthur, who won the Medal of Honor at Missionary Ridge, Tennessee, in 1863 as a Union officer. Douglas grew up on military posts with a single aim in life: to emulate his famous father by also winning the Medal of Honor, the nation's highest decoration for bravery. 'My first memory was the sound of bugles,' he would later recall. 'I learned to ride and shoot even before I could read or write – indeed, almost before I could walk or talk.'

MacArthur entered West Point in 1898 and not only achieved the distinction of first captain of the corps of cadets but also graduated first in the class of 1903, a glittering academic record only surpassed by two other graduates, one of whom was Robert E. Lee.

After commissioning as an engineer officer, he was posted to the Philippines, where he soon had the first of many brushes with death, waylaid by two desperadoes while on a surveying mission in the jungle. MacArthur later wrote, 'Like all frontiersmen, I was expert with a pistol. I dropped them both dead in their tracks.'

Back in the United States, MacArthur served as an aide in the White House to President Theodore Roosevelt and in 1914 participated in the expedition to Vera Cruz, Mexico, where he was nominated for, but not awarded, the Medal of Honor for his exploits while on a reconnaissance mission. He had gone behind Mexican lines and daringly managed to purloin three locomotives before returning with them to friendly lines through a hail of gunfire.

First World War service

By the time the US entered the First World War in 1917, MacArthur had achieved the rank of full colonel. Sent to France as chief of staff of the 42nd 'Rainbow ' Division, a New York National Guard unit, he served in the front lines as a brigade commander. Utterly fearless and constantly exposing himself to enemy fire, MacArthur earned exceptional distinction as the most highly decorated American soldier of the war, winning two Distinguished Service Crosses (the second highest decoration awarded for bravery), seven Silver Stars, the Distinguished Service Medal and two Purple Hearts for wounds received in battle. MacArthur was also recommended for the Medal of Honor for leading his brigade during the battle for the Côte de Châtillon in the Meuse–Argonne campaign in October 1918. However, the commander-in-chief of the American Expeditionary Force (AEF), General John J. Pershing, inexplicably denied it. By the end of the war, MacArthur was a brigadier general in command of the division during the Sedan offensive. Secretary of War Newton D. Baker called MacArthur 'the greatest American field commander produced by the war'.

The inter-war years

MacArthur emerged from the First World War with an impressive service record and eight rows of ribbons on his uniform. Whereas most other officers were demoted in the peacetime army, not only did MacArthur keep his rank of brigadier general, but his reputation as one of the army's most brilliant officers earned him the prestigious appointment of superintendent of West Point. During his tenure from 1919 to 1922, he overcame the entrenched status quo by instituting badly needed reforms that modernized the academy's archaic academic curriculum and put the brakes on its lax discipline and notorious 'hazing' practices of ritualized abuse of new recruits.

During the 1920s MacArthur served two tours of duty in the Philippines, one of them in command of the Philippine Department, and in 1925 he became the army's youngest major-general. His reputation for supporting amateur athletics led to his appointment to chair the US committee for the 1928 Olympic Games.

In 1930 President Herbert Hoover appointed MacArthur army chief of staff. His most difficult undertaking was merely to hold the small peacetime army together during the Great Depression and stave off repeated attempts to decimate its annual parsimonious budget appropriations. He also reformed the army education system and the results paid off handsomely when war came and men who had benefited from MacArthur's initiatives were given important commands; they uniformly succeeded.

In 1932 MacArthur was at the centre of the controversy over the infamous Bonus March on Washington by thousands of destitute First World War veterans who were protesting the government's failure to make good on its promise of bonus money. With the full support of MacArthur, who acted with unseemly relish, the army was ordered to disperse the marchers and the resulting violence – which included the use of tear gas, sabres and bayonets – marked one of the most shameful episodes of American history.

He imprudently claimed to have saved the nation from 'incipient revolution' by a mob of 'insurrectionists'; the long-term damage from the Bonus March was incalculable in an era where the military was already weakened and under fire, and facing still more budgetary cutbacks. MacArthur and the army became the public exemplars of an ungrateful nation that rewarded its veterans by gassing, bayoneting and shooting them. During his long and distinguished military career Douglas MacArthur was at the heart of numerous controversies but none did more to tarnish his reputation permanently, and that of the army he headed, than his actions over the Bonus March.

Although he embraced President Franklin D. Roosevelt's New Deal, MacArthur became unpopular with the president for his vocal opposition to pacifism and America's rampant isolationism. When his term as chief of staff ended in October 1935, Philippine president Manuel Quezon, an old friend from his earlier duty, persuaded MacArthur to come to Manila to help form and train the Philippine army. Offered the opportunity to return to a place he dearly loved, he accepted at once. It never seems to have dawned on the politically naïve MacArthur, who was thought to be considering a run for president, that it suited both Roosevelt and his many other enemies to have him 11,000 miles from Washington.

Those who knew or served under MacArthur either admired his genius or despised his narcissism. One of his principal assistants in Washington and the Philippines in the 1930s was Dwight Eisenhower, who has said of him that: 'He did have a hell of an intellect! My God, but he was smart. He had a brain.'

In Manila, MacArthur's ego was on full display. He officially retired from the army in 1937 in order to accept an appointment as a field marshal in the Philippine army, the only American officer ever to hold such a rank. The appointment was widely seen as dubious and was derided by Eisenhower, who was far from alone in his belief that not only was MacArthur being disloyal to the army, but that it was absurd for him to be the field marshal of a virtually non-existent force.

In large part, MacArthur was a victim of the Great Depression during his tenure in Manila. His constant feuds with Washington over funding and equipment for the fledgling Philippine army were equally harmful. Grossly inadequate funds and insufficient military equipment left MacArthur unable to fulfil his mission of creating a viable Philippine army. But he would spend much of the Second World World preoccupied with the Philippines, albeit as commander of US Far East forces.

The Philippines in the Second World War

With war clouds gathering in the Pacific, Roosevelt recalled MacArthur to active duty in the summer of 1941 and named him commander of US Far East forces, sending him back to the Philippines. When Japan attacked Pearl Harbor on 7 December 1941, MacArthur again became a four-star general. His years of attempting to forge a defence of the Philippines came to naught in the succeeding days, the islands being indefensible and the Philippine army no match for the better-trained and far more powerful Japanese.

Japanese aircraft attacked Clark Field near Manila on 8 December, where the US B-17 bomber fleet was caught on the ground and severely mauled. This was followed by Japanese landings on the island of Luzon that, by the end of December, proved unstoppable. Manila was evacuated and by early January 1942 US and Philippine forces retreated to the Bataan peninsula and the island fortress of Corregidor. During the first desperate months of 1942, the fate of MacArthur's force was sealed and by early March, with surrender only a matter of time, Roosevelt ordered MacArthur to Australia. He was secretly evacuated from Corregidor by PT boat on 11 March and after a harrowing escape vowed, 'I shall return'. Despite the loss of the Philippines MacArthur was awarded the Medal of Honor, largely to help boost morale at home and to counter Axis propaganda that portrayed him as a coward who had deserted his troops.

Indeed, many never forgave him for leaving behind his embattled force led by Lieutenant General Jonathan Wainwright, which held out until 7 May before surrendering. The survivors became part of the infamous Bataan Death March. The loss of his Philippine force was anguish for MacArthur, who declared that: 'Through the bloody haze of its last reverberating shot I shall always seem to see a vision of grim, gaunt, ghastly men still unafraid.' Although unwarranted, the derisive term 'Dug-out Doug' forever tainted MacArthur's reputation.

Although it was impossible to have reinforced the Philippines, MacArthur was unforgiving and bitterly blamed Washington for the loss of his force there. His relations with Roosevelt ranged from the frosty to the downright disrespectful.

Appointed supreme commander of Allied forces in the Southwest Pacific (SWPA), MacArthur established his headquarters in Australia and initiated campaigns to recapture key islands, beginning with New Guinea in 1943. He also fought a bitter turf battle with the US Navy over control of the war in the Pacific. MacArthur wanted to focus on recapturing New Guinea and the Philippines, while the admirals favoured a strategy of island-hopping across the central Pacific in a manner that made better use of the US carrier fleet. In the end, both strategies were carried out. MacArthur was especially effective in his employment of air power in support of ground operations.

On 20 October 1944 MacArthur carried out his promise to return when he waded ashore on the island of Leyte during the successful campaign to recapture the Philippines (see also pp. 307–9). After retaking the islands, and as the war neared its end, he began to plan for the invasion of Japan in late 1945. However, the employment of the atomic bomb on Hiroshima and Nagasaki brought about Japan's unconditional surrender, which was taken by a triumphant MacArthur on the deck of the battleship USS *Missouri* on 2 September 1945.

Proconsul of Japan

Designated the Supreme Commander of Allied Powers, MacArthur now took on perhaps his most demanding assignment yet: the governance of a defeated, shattered nation. Determined to bring about a peaceful and orderly transition to eventual independence, he became an administrator and under his even-handed tutelage Japan not only recovered but made the transition from dictatorship to democracy and, eventually, to the status of a world economic power. For five and a half years MacArthur was literally the governor of Japan, a benevolent but firm-handed leader who guided, cajoled, and where necessary compelled a series of actions that brought about reforms, massive reconstruction, and the formation of a democratic state.

MacArthur also believed that it was essential for the morale and the future of the nation that Emperor Hirohito be kept as Japan's spiritual leader and not punished or forced to abdicate for the acts of the militarists who brought about the war. MacArthur's enduring legacy in Japan was the creation of a constitution that is still the basis of a modern democratic nation.

War in Korea

On 25 June 1950, in the divided Korean peninsula, the army of North Korea crossed the border, soon overran Seoul and was on the verge of conquering all of South Korea. US occupation forces were unprepared, largely untrained, and in disarray from the surprise invasion. The United

Nations Security Council authorized the creation of a UN armed force to aid South Korea. MacArthur was named supreme commander and immediately began planning an amphibious invasion behind enemy lines at Inchon that was brilliantly carried out on 15 September 1950.

The Inchon landings were a masterstroke. MacArthur's daring gambit caught the North Koreans by surprise and within a matter of days the capital of Seoul was liberated and the North Korean Army was in full retreat, pursued by UN forces deep into North Korea.

Macarthur ignored warnings from Beijing that such an advance was deemed provocative and would bring China into the war. In November 1950 he committed the greatest blunder of his career when he boldly announced that the war was won and his troops would be home by Christmas, recklessly dismissing the mounting, incontrovertible intelligence that the Chinese People's Liberation Army was massed in North Korea. By pursuing the North Korean army to the Yalu River on the Manchurian border, his UN force became dangerously exposed. The result was a humiliating and costly fighting retreat from 'the frozen Chosin' and an emergency evacuation from the port of Hungnam.

MacArthur publicly and privately disagreed with the American policy of limiting the war in Korea and avoiding a possible larger conflict with China. His blatant challenge to civilian authority and his insolent behaviour toward his commander-in-chief led President Harry S. Truman to relieve him of command for insubordination on 11 April 1951. MacArthur's sacking was enormously controversial, with Truman cast as a villain for removing one of America's greatest military heroes from command. When MacArthur returned to the United States he was greeted by an outpouring of public adulation and invited to address a joint session of Congress.

His compelling speech was broadcast to the nation on both radio and television. MacArthur insisted that in the twilight of his life he was only doing his duty and that from the time he first entered West Point he had served his country for fifty-two years. He concluded by quoting an old barrack-room ballad: '"Old soldiers never die, they just fade away." And like the old soldier of that ballad, I now close my military career and just fade away – an old soldier who tried to do his duty as God gave him the light to see that duty.' The enormous controversy notwithstanding, Macarthur had indeed not only mishandled the war but also violated the principle that soldiers do not dictate foreign policy. Truman's courageous decision was wholly justified and eventually vindicated by a Congressional investigation. MacArthur's strong defence of his actions in testimony before Congress was discredited by a parade of distinguished witnesses, including Secretary of State George C. Marshall and General Omar N.

Bradley, the chairman of the joint chiefs of staff, both of whom persuasively argued that his actions would have pushed the United States into a ruinous and unnecessary war with Red China, and possibly with the Soviet Union as well.

MacArthur's legacy

MacArthur was a man of many contradictions. As one of his biographers has written: 'MacArthur's life was shaped by the nineteenth-century belief that history is created by the actions of great men ... His only goal in life was to be remembered with the great.'

One who knew him well in the Philippines offered this thoughtful assessment: 'There was never any middle ground; people either idolized this man or hated him, while all the time he dwelt on another mental plane, and was probably seldom aware of either the worship or the hate.' His career-long penchant for surrounding himself with sycophants who told him only what he wanted to hear was his undoing in Korea. The deliberate suppression of the incontrovertible fact of the Red Army's presence in Korea by his chief intelligence officer, Major General Charles A. Willoughby, was the most costly example.

MacArthur was thought to have political ambitions in the 1930s, which is a principal reason why Roosevelt orchestrated his exile to the Philippines. In 1952 it was widely believed that he might attempt to run for president, but public adulation quickly faded and he did not challenge his former subordinate, Dwight Eisenhower, who won the Republican nomination – and the election.

Throughout his life, MacArthur never forgot what West Point had taught him. His most famous and moving speech occurred when he returned in 1962 to address the Corps of Cadets. He reminded them that: 'The soldier, above all other people, prays for peace, for he must suffer and bear the deepest wounds and scars of war. But always in our ears ring the ominous words of Plato, that wisest of all philosophers, "Only the dead have seen the end of war" ... in the evening of my memory, always I come back to West Point.' Always there were echoes and re-echoes of Duty, Honour, Country. 'Today marks my final roll call with you, but I want you to know that when I cross the river my last conscious thought will be the Corps – and the Corps – and the Corps. I bid you farewell.'

To his detriment, MacArthur's legendary vanity and penchant for self-promotion has left him just as controversial today as he was at the time of his relief from command by Truman in 1951. It has also tended to obscure his exceptional accomplishments. He was perhaps the most dynamic general in modern American history, and while his legacy and his reputation continue to be contentious, what can never be taken away from Douglas MacArthur was a life of great and courageous adventure. As he had predicted, MacArthur did fade away and his death in April 1964 at the age of 84 ended a brilliant but flawed military career.

WILLIAM SLIM

1891–1970

HUGO SLIM

BILL SLIM was one of history's best all-round commanders. He had a genius for planning, strategy, logistics, manoeuvre and morale, all of which he combined to extraordinary effect against Japanese forces in one of the hardest theatres of modern war – the mountains, plains and jungles of eastern India and Burma. Learning fast from his initial defeat during the Japanese invasion of Burma in January 1942, Slim pioneered much of the modern practice of warfare by combining mobility and concentration of force on the ground with sophisticated air supply over an enormous area – one that had few roads and which was drenched by torrential monsoon rains.

Slim's army of 1 million men was the biggest army fielded by Britain and its empire during the Second World War. It was a truly multiracial army. Only 12 per cent of Slim's troops were British. The great majority were from India, Nepal and East and West Africa. Slim's rapport with his troops – to whom he was affectionately known as 'Uncle Bill' – was unique among the commanders of the 1939–45 war. Always dressed in his Indian army battle dress, with Gurkha hat, binoculars and a machine-gun slung over his shoulder, Slim spent much of his time visiting and encouraging his men in the terrible conditions in which they fought. As their leader, he was calm, direct, humorous and determined to give them what they needed to do the job. Slim's striking confidence, clear thinking and sense of purpose were embodied in his jutting chin; but he was also personally modest, and possessed of a genuine capacity for self-criticism – a rare quality in the egos of great men. He claimed little for himself, but always recognized the brilliance of his staff and the skill and courage of his men.

From Birmingham to Burma

William Slim was born in Bristol in 1891, the younger of two sons in a struggling middle-class family. In 1903 his father's business failed and the family moved to Birmingham, where Slim's devout mother ensured him a good Catholic education at St Phillip's Grammar School, followed briefly by entry into Birmingham's famous King Edward's School. But his education was cut short at 16 when his father's business failed again, and all the family's resources were put towards his brother's medical degree. Slim now had to earn his living. He worked first as a primary-school teacher in a poor part of Birmingham for two years, and then for four years as a clerk in an ironworks. Here he got to know factory life and working people.

Slim had always dreamt of a military career, but a commission was never within his means. He did, however, somehow manage to join the Officer Training Corps at Birmingham University, even though he was not a student there. When war broke out in 1914 he was gazetted as a second lieutenant in the Royal Warwicks, with whom he served at Gallipoli – where he was badly wounded – and then in Mesopotamia, where he was awarded the Military Cross and wounded once more.

Transferred to the Indian army in 1919 – where commissions were cheaper – Slim joined the Gurkhas in 1920. He managed to support his military career with lucrative short-story writing for London magazines, using the pen name William Mils (Slim backwards). In the Indian army Slim stood out as the leading intellectual soldier of his generation – first as a pupil and then as a teacher at staff colleges in Quetta, Camberley and Belgaum – while also seeing regular active service on the North-West Frontier.

With the outbreak of the Second World War, Slim was

given his first brigade command, in Sudan and Eritrea, where he was wounded again fighting the Italians. This was followed by his first divisional command, in Iraq, where he fought the Vichy French and their allies before going on to invade Persia. In March 1942 Slim was chosen as the commander of Burma Corps, under General Alexander. His task was to stop the Japanese, who had invaded Burma, and were set on advancing on India.

More with less

In the three years of the Burma campaign (1942–5) Slim was tested, and proved himself a master, in all four of the great challenges of command – retreat, regrouping, defence and offence – usually with less equipment and resources than his peers in other theatres.

In his leadership of Burma Corps, Slim oversaw the longest retreat in British military history, some 900 miles. Poorly trained, ill-equipped and constantly compromised by Alexander's political efforts to keep on good terms with the Chinese, Burma Corps was defeated and retreated out of Burma in May 1942. During the retreat Slim kept looking for any chance to attack, while gradually withdrawing his beaten forces and avoiding a rout. Arriving defeated, emaciated but in order in India, his men cheered him at his farewell address to them. Slim later commented that 'To be cheered by troops whom you have led to victory is grand and exhilarating. To be cheered by the gaunt remnants of those whom you have led only in defeat, withdrawal and disaster is infinitely moving – and humbling.'

In June 1942 Slim was made commander of XV Corps, part of General Noel Irwin's Eastern Army focused on the Arakan in the south. It was in XV Corps that Slim began to implement his new tactics and training for beating the Japanese. Irwin sidelined Slim, however, rejecting his strategy and insisting on taking operational control of the first Arakan campaign himself. Irwin's conventional and unimaginative offensive tactics in the Arakan produced disaster. Irwin tried to blame these new defeats on Slim, who once again had to save the situation by resuming command for an orderly retreat. At the end of May 1943 Irwin was relieved of his command and replaced by General George Giffard, with whom Slim had an excellent working relationship.

Giffard ordered Slim to prepare for a new offensive in the Arakan. This time Slim had the monsoon season to train XV Corps properly and the freedom to design his own strategy and tactics. In October 1943, with the arrival of Lord Louis Mountbatten as Supreme Allied Commander Southeast Asia, Giffard was made commander of land forces and Slim was appointed as commander of the new Fourteenth Army to replace Eastern Army. Slim's training and organization of the Fourteenth Army (whose badge he designed himself) is the classic example of the making of an army. In forging this new force, Slim shaped the spirit and confidence needed to win, while developing the tactics and training necessary to fight in extreme and often isolated surroundings. He also procured or improvised the right equipment, and built up a winning morale throughout his diverse force. The results were spectacular.

In late 1943 Slim asked XV Corps, now under General Philip Christison, to deliver Operation Cudgel in the Arakan. After a slow start, XV Corps began to use Slim's new approach – treating Japanese tactics of encirclement and infiltration as an opportunity not a threat. Aggressive patrolling was combined with Slim's famous 'boxes'. Whenever units were encircled or attacked, they were to stand firm and fight aggressively. Supplied by air, and with all non-combatant support staff trained to fight as well, they could then defeat relentless waves of Japanese attacks. Learning from the Japanese 'hooks', Slim also encouraged XV Corps to avoid frontal assaults, infiltrate behind Japanese positions and attack them in their rear or on their flanks, so turning the Japanese out of heavily defended positions to ground of the Fourteenth Army's choosing. XV Corps did all this and won. It was the first defeat of Japanese forces and was a tonic for the confidence of the Fourteenth Army. But it was only the beginning.

Slim was now making ready for the major Japanese

offensive against India, which, he guessed rightly, would come further north, in Assam. In his planning, preparations and mobility, Slim showed himself to be a master of aggressive defence. Ignoring the advice of others to take the battle to the Japanese by moving into Burma, Slim preferred to wait for them where he was strongest and they would be weakest. As he expected, Mutaguchi's Fifteenth Army came at him in Assam in Operation U-Go (March 1944), which was intended to spearhead the invasion of India. Slim's plan was to lure Mutaguchi on to the plain around Imphal and to smash his army there.

Imphal and Kohima

Between March and July 1944 Slim led one of the fiercest and most complex battles of the war around Imphal and Kohima. The fighting lasted for five months. Much of it was face-to-face, reminiscent of the Somme in the First World War, as the Japanese continuously fought to the death. Slim's approach was subtle and involved deliberate fighting withdrawals by two divisions to pull the Japanese army on to the plain. His plan involved the anticipated siege of Imphal and the pre-arranged and superbly organized movement of two divisions by air across a vast and difficult front.

But there were also misjudgements, to which Slim typically confessed. He got his timing wrong on the withdrawal from Tiddim, so that the 17th Division had to fight for their lives as they withdrew. He was also surprised when Mutaguchi attacked with a division at Kohima when Slim had expected him to strike instead at the Fourteenth Army's main supply base at Dimapur. Extraordinary courage in the defence of Kohima saved the day, and famously involved repeated hand-to-hand fighting across the district commissioner's tennis court – the white lines of which remain marked out in cement today as the centrepiece of the Allied cemetery in this beautiful mountain town.

But Slim's plan worked. Mutaguchi met ferocity, professionalism and mobility he had never expected. As Slim anticipated, this infuriated him and he refused to give up, repeatedly throwing his troops at Slim's forces, whose new confidence now saw every Japanese attack as an opportunity to kill more of the enemy. When the Japanese did eventually retreat, the Fourteenth Army were aggressive and thorough in their pursuit. Mutaguchi's army was decimated. From 105,000 men, the Japanese suffered 90,000 casualties, of whom 65,000 were killed. Only 600 surrendered. Imphal was the biggest ever defeat in Japanese military history, and it turned the tide. After the battle, Slim and his three corps commanders, Christison, Geoffrey Scoones and Montagu Stopford, were all knighted by Viscount Wavell, viceroy of India, on a dusty bulldozed field at Imphal.

The reconquest of Burma

Slim now led the Fourteenth Army in a dashing reconquest of Burma itself (January–May 1945). Although sea- and airborne operations were ruled out – the invasion of France had taken all available resources – Slim was convinced that his men could do it. But first he had to ensure that they trained for a more open fighting style suited to the plains of central Burma, which gave good opportunities for massed armour, long-range artillery and mechanized infantry. The reconquest also meant still more demanding preparations in logistics and engineering, as the advance involved ever longer lines of communications, the crossing of the Irrawaddy, one of the world's biggest rivers, and the usual race against the monsoon. But it would be made easier by the Allies' complete aerial superiority. Even as he went full tilt on the offensive, Slim still remained thoughtful and well prepared.

As a strategist, Slim was always cunning – reading, luring and deceiving his enemy while manoeuvring himself into a position where he could beat him outright. In so doing, he frequently turned the tactics of the Japanese against themselves. In the offensive to retake Burma – Operation Capital – Slim brought this approach to a crescendo by secretly moving a whole corps in a 320-mile silent march down dirt tracks to attack the Japanese rear at Meiktila – a move that his defeated opponent, General Kimura, later described as Slim's 'masterpiece'.

Meiktila

Slim's initial plan for the decisive battle in central Burma was to meet the Japanese after crossing the Irrawaddy and defeat them across a broad mobile front. This too was what his opponent expected and for which he also began to prepare. But when Slim realized that he and his enemy were both preparing for the same thing, he typically decided to surprise him by luring him one way and then also hitting him hard somewhere unexpected. So Slim added Operation Extended Capital to his original plan.

Extended Capital was a necessarily covert, high-risk and daring plan, which set out to surpass the Japanese at their own game by launching a long and powerful hook south behind their main lines to strike and take the main Japanese supply and administrative base at Meiktila. At a stroke, this would cripple Japanese supplies, force their army to turn and fight in two directions at once and also provide Slim with a classic hammer-and-anvil situation – crushing Japanese forces between Fourteenth Army divisions fighting from both north and south.

Slim entrusted the 320-mile silent march south to IV Corps under Messervy, and while it moved, he effectively deceived his opponent by setting up a false IV Corps HQ, which sent and received fake radio signals for the weeks necessary for Messervy to outflank the Japanese. Meanwhile, Stopford led XXXIII Corps in the attack on Mandalay and Christison led XV Corps in amphibious assaults along the Arakan coast, capturing the airfields needed to supply the Fourteenth Army as it moved south.

The attack on Meiktila by IV Corps began on 1 March 1945, and the town was taken four days later. Cowan and others went on to conduct aggressive defensive fighting for a further month as the Japanese, who had been totally surprised, fought hard to regain and secure their rear. But it worked. Slim had lured General Honda, the Japanese army commander, into a battle of destruction around Meiktila simultaneous with that around Mandalay, while the amphibious operations in the south had made possible constant air supply.

Slim's extraordinary deception, together with the simultaneous and interlinked advances of the three corps with their air and naval support, are widely recognized as one of the finest feats of combined arms during the Second World War. The Japanese army was steadily crushed and dispersed, while the Fourteenth Army raced on towards Rangoon, which was initially taken from the sea, after all, on 3 May after IV Corps had been held up by the early arrival of the monsoon.

Slim as a leader

The Fourteenth Army was mainly Indian; only one in eight of its men were British. Other vital contingents came from Nepal, East and West Africa, Australia, New Zealand and the USA, plus Chinese divisions under General Stilwell's command in northern Burma. Fluent in Hindi, Urdu and Ghurkali, Slim was able to communicate intimately in the three common languages of his troops. He was a master of morale, with a natural, relaxed and unaffected genius for relating and communicating with his men. He understood their fears, their needs and their hopes, realizing instinctively that fighting and leadership are emotional as well as physical and intellectual endeavours. He held his army in great affection, an affection that his army returned in equal measure. Robert Lyman and others who have written about Slim see this emotional connection as the hallmark of Slim's generalship – that he was truly loved by his men in a way that had not been seen in the British military since Nelson and Wellington.

But there was more to Slim than just charisma. He was also a rigorous and imaginative planner. His men loved him because he also delivered. Slim's genius for logistical and administrative organization, together with the exceptional abilities of his senior administrative and engineering staff like Alf Snelling and William Hasted, meant that the Fourteenth Army could support divisions on the move in the most difficult of terrain, even during the monsoon. This was done by extraordinary and well-planned feats of engineering, medical care and air support. Slim was sometimes irritated by comments that his army was famous for 'improvising'. Yes, it constantly innovated and made use of whatever was available, but it was always, he insisted, working to detailed plans and objectives that were well prepared in advance. To cross the Irrawaddy – a river three times as wide as the Rhine – Slim's engineers had to build the ships and the bridges they needed themselves, using teams of 3,000 elephants to haul the thousands of trees required. Where there was no parachute silk for air supply, they used local jute to make 'parajutes'. They grew their own food in massive farms, and baked millions of bricks to make their own roads. They also pioneered treatments for malaria, dysentery, typhus and other diseases, with considerable success: in 1943 for every man evacuated wounded 120 were evacuated sick; astonishingly, by 1944, in the unhealthiest theatre, the Fourteenth Army was the healthiest army in the world, with only one soldier in every thousand sick each day.

Much of this planning and efficiency involved intimate combined land–air operations. With Slim's vision, and gifted colleagues like Jack Baldwin and Stanley Vincent in the RAF and George E. Stratemeyer in the US Air Force, the battle for Burma was won by unprecedented cooperation between army and air forces, a relationship that Slim described as 'a close brotherhood'. By 1945 nearly 90 per cent of the Fourteenth Army's supplies were provided by air, in some 7,000 sorties each day. The gradual increase of this air support – which was years ahead of its time – was made possible by the energetic endorsement of the Supreme Allied Commander, Mountbatten, with whom Slim shared a common vision and enjoyed a highly creative working relationship.

Slim's relationships with senior officers were either very good or strangely problematic. His own corps and divisional commanders respected him enormously for his relaxed style and his calm ability to delegate, empower, discuss and actively support. Many, but not all, were long-standing Indian army friends and colleagues. With Mountbatten, Giffard and the famously difficult American, General Stilwell, Slim got on extremely well. With Irwin and Oliver Leese things were difficult. Some of these difficulties may have turned on the old British hang-up about class. Slim, the grammar-school boy from the Indian army, may have been looked down upon or unconsciously underrated by some from the smart upper-class set of the British army. This, in turn, may well have raised Slim's hackles and elicited a bit of Brummie scorn. Yet, in any clash with senior officers, it was Slim who was vindicated and twice (with Irwin and Leese) he famously took the jobs of those who tried to sack him. With the maverick Orde Wingate, Slim was patient but firm, always rightly convinced that only a whole-army approach would defeat the Japanese. Slim realized that while Wingate's behind-the-lines guerrilla operations with his Chindits could be useful, their effect was overblown in the imagination of their leader and his prime ministerial patron in far off Downing Street.

The battles in which Slim led his troops were fierce, fought in the harshest climate with disease ever present, and against an unrelenting enemy who always fought to the death. The campaign saw great bravery: the Fourteenth Army numbered among its men twenty-nine recipients of the Victoria Cross. In these extreme conditions, Slim forged an extraordinary multiracial force, made daring plans, overcame huge logistical obstacles and called forth immense willpower, courage, talent and good spirit in his troops. Slim always knew that it was his men who 'did it', but many of them felt that his leadership had carried them beyond themselves and that, even if they had made the victory, Slim had somehow made them.

Later life

After the war, Slim initially retired from the army and became vice chairman of the newly nationalized British Railways. But he was soon recalled from civilian life, made a field marshal and served as chief of the Imperial General

Staff from 1948 to 1952, the first Indian army officer to be appointed to the top job in the British armed forces. From 1952 to 1960 Slim was a highly successful governor-general of Australia. During this time, he wrote his famous book about the war in Burma, *Defeat into Victory,* which is probably the finest commander's account of a campaign since Julius Caesar's *Gallic Wars. Defeat into Victory* is still studied in military academies all around the world, a beautifully written and modest masterclass in leadership and military command. In 1963, towards the end of his life, Slim was made governor of Windsor Castle. Finally, in 1970, he retired to London and died in December the same year.

In the many busy years of his life, Slim was supported throughout by a remarkable woman from Scotland, Aileen Robertson. They had met on a ship to India in 1924 and married the following year.

GERALD TEMPLER

1898–1979

JOHN HUGHES-WILSON

FIELD MARSHAL TEMPLER was one of the most successful British military commanders of the twentieth century. Most soldiers spend their careers awaiting the call to arms, but Templer was an exception. He joined the army in the middle of the Great War, fought to put down colonial rebellions in the inter-war years, saw the Second World War out from start to finish and then went on to fight and defeat the communist insurgency in Malaya in the twilight of the British Empire. He could with justification claim to be one of Britain's most successful generals; his Malayan campaign is a model for counter-insurgency operations.

Gerald Walter Robert Templer was one of the remarkable crop of soldiers grown out of the martial soil of Ulster during the twentieth century, along with Alexander, Alanbrooke, Auchinleck, Dill and Montgomery. Born in 1898 as the son of a serving officer, the combination of Wellington College and Sandhurst seems almost pre-ordained. In November 1917 he joined his father's old regiment, the Royal Irish Fusiliers, on the Western Front, taking part in the retreat from the Somme in spring 1918 and then advancing as part of Haig's victorious last offensive in the autumn.

In 1919 he was in the Caucasus as part of the Allies' abortive attempt to overthrow the Bolshevik revolution in Russia. 'All great fun,' he wrote, 'but pretty unsatisfactory from the political point of view.' Later, during the Cold War, he claimed that he was the only serving senior officer who had actually fought the Red Army. By 1922 Lieutenant Templer was back in England to become bayonet-fighting champion of the army and part of the British Olympic hurdling team of 1924.

In 1927 he went to the army's staff college at Camberley. His first staff job, at HQ 3rd Division, was a disaster. He clashed with his senior staff officer, who gave him an adverse report and even recommended that he be dismissed from the army. Templer protested, and appealed to his general who ripped up the offending report and saved his career by packing him off to Northern Command. By 1935 Captain Templer was hunting down terrorists in Palestine as a company commander in the Loyal Regiment.

For his efforts Templer was awarded a DSO (Distinguished Service Order), an unusual honour for a junior commander. Like his contemporary Orde Wingate, Templer learned a lot about his trade among the rocky hills of Judea and later admitted to actually weeping over the tragedy of the Arab–Jewish problem. But he always claimed that it was 'Palestine that taught [me] the mind and method of the guerrillas'.

The outbreak of Hitler's war saw him as a staff officer in the Directorate of Military Intelligence, where he helped to lay the foundations of two successful organizations, MI Special Operations (later to become SOE) and the Army's Intelligence Corps. This experience gained him the post of GSO 1 (Int) in the British Expeditionary Force (BEF) in France in 1939 and later as chief of staff to the ill-fated 'MacForce' which tried to keep the western corridor of the retreat to Dunkirk open before the Wehrmacht's rampaging panzers. On 27 May 1940 Lieutenant Colonel Templer was evacuated from Dunkirk; one of the many British officers who discovered that defeat is the hardest school of all.

From soldier to general

If his career until then had made him a competent soldier, the next five years saw the making of Templer as a general. From a brevet lieutenant colonel in 1939 he rose to a general in 1945. He was appointed brigadier commanding 210 Brigade in November 1940, went as the chief of staff

to V Corps in May 1941, and by April 1942 was a major-general commanding 47th London Division. This was meteoric, and his appointment as corps commander of Second Corps at Newmarket in the autumn of 1942 astonished many. He was, at 44, the youngest lieutenant general in the army. He was renowned as a martinet with high standards and drove himself and his men hard.
Like Wellington before him, he had an eagle-eye for sloppiness, inefficiency and slackness, and would not tolerate those who failed to share his commitment and drive. Perhaps his greatest quality as a general was his ability to focus on the mission: that rare ability of great leaders to distil the selection and maintenance of the aim. Templer always knew what he wanted and he always knew how best to get it.

Templer then did an extraordinary thing: he demanded a reduction in rank and a posting to the battlefront – anywhere. With a true soldier's instinct he knew that he had to head for the sound of the guns. A bemused War Office granted his wish and by the summer of 1943 *Major*-General Templer was in the North African theatre commanding the 1st Division. By October he had been switched to lead 56th Division in Italy as the Allies struggled across winter mountains and rivers in the face of determined German resistance. In November his division assaulted across the River Garigliano to evict the Germans from Monte Cassino. The attack failed with heavy casualties.

In January 1944 the Allies tried again, this time with an overwhelming array of firepower, and blew the Germans – and Monte Cassino – off the hill for good. Hardly had Templer's division settled in for an advance on Rome when they were whipped away to reinforce the hard-pressed Anzio beachhead. At one point he personally ordered every staff officer, cook and sanitary orderly into the firing line to beat off a German assault. In March 1944 his exhausted division was pulled back to Egypt to refit, but by 26 July Templer was again back in Italy, this time to command 6th Armoured Division.

An unusual wound

Templer was a hands-on, front-line general, and blacked-up infantry patrols would get back to their lines to find an impatient GOC himself leading their debriefing. 'He was like a red-hot poker,' said one officer. The fiery general's war ended when he joined the ranks of his soldiers wounded in action. On 10 August he was driving up to the front when a truck pulled off the road and hit a landmine on the verge. The vehicle exploded and a looted grand piano fell on to Templer's vehicle, breaking his back. Seriously injured, he was taken back to Britain and was not ready for action again until the spring of 1945. He would say ruefully that he 'was the only British officer to be wounded by a piano'.

In the last months of the war he was seconded to convalesce at SOE, where he was one of the architects of Operation Foxley, a plan to assassinate Hitler with a sniper at his mountain retreat, the Berghof. Other counsels prevailed, however. Churchill decided that Hitler alive as a rotten strategist was much more useful to the Allies than Hitler as a dead Nazi martyr.

The post-war years

The war's end saw Templer take over a challenging appointment as military governor of the British zone of occupied Germany. He announced his policy with brutal clarity: 'I intend to be firm to the point of ruthlessness ... I still have to meet a German who says he is sorry.' In late 1945 he sacked the post-war mayor of Cologne for laziness and incompetence. This was typical Templer and certainly galvanized the other builders of the new West Germany. However, as the mayor in question was Konrad Adenauer, who went on to become the first chancellor of the new Bundesrepublik Deutschland, the decision came back to haunt him. To his credit, Adenauer never bore Templer a grudge and sent him a present of a crate of wine every time he came to London.

War Office appointments in cash-strapped Whitehall

taught Templer the need to get along with politicians. Finally in 1950 he was put out to grass as GOC Eastern Command. There he might have remained if terrorists in Malaya had not ambushed and shot the governor-general, Sir Henry Gurney. Whitehall suddenly realized that the insurgency and civil war 'emergency' (so-called for insurance reasons) was getting out of hand, and decided to appoint a senior soldier with, unusually, both civil and military powers.

At the Rideau Hall conference in Ottawa in January 1952, discussing the weakness of Commonwealth defences, Winston Churchill looked down the dinner table. 'Templer,' the great man growled through the cigar smoke, waving a brandy glass. 'Malaya!' he bellowed. 'Full powers, now, Templer. Full powers,' he added. About ten minutes later, after a whispered confab with his startled advisers, Churchill broke through the conversation again. 'Full power, Templer. Very heady stuff. Make sure you use it sparingly.' Thus was Sir Gerald Templer appointed governor-general of strife-torn Malaya, with greater military and political powers than any British soldier since Cromwell.

A high commissioner at war

Templer arrived in Malaya in February 1952, by which time it looked as if he might be too late. The post-war communist uprising against British rule was succeeding. Over 250,000 soldiers and policemen were combing the jungle fruitlessly trying to locate and destroy a few thousand terrorists led by Chin Peng. Ironically it was the British who had armed and trained the guerrillas in the first place and used them in their undercover war against the occupying Japanese; they had even given Chin Peng an OBE. Now the 'CTs' ('Communist Terrorists') had turned on their colonial masters. Templer's orders from Churchill were clear: smash the communists and turn Malaya into a single, self-governing, democratic state.

The political challenge would take time but Templer quickly saw that the real problem was a lack of coordination at every level. The army and the police did not work together; the colonial planters and civil service acted as if the insurgency was the military's problem; and there was deep suspicion between the Chinese and the Malay populations. Morale was low. With characteristic energy Templer decided to shake up the situation from the start. He travelled by armoured car or helicopter and suddenly the spare figure of the sharp-eyed boss was everywhere. Suddenly he, and not the CTs, was driving events. The sleepy civil service was shaken to its core. On one occasion the new governor-general asked a startled civil servant what he did, to be told, 'Nuclear emergency planning, sir.' Templer considered this and said, 'What's the likelihood of a nuclear strike on Malaya?' The bureaucrat laughed, 'Zero, I would say.' 'Good,' rasped Templer. 'I agree. You're fired.'

The whiplash of Templer's tongue and his eagle-eye settled on the slack, the inefficient and the lazy. Even colonial rubber-plantation managers in the middle of the jungle suddenly realized that they were part of the war. Templer listened to one planter moaning about inadequate army and police protection. 'Do you ever go down and talk to the troops or the police?' he demanded. 'Of course not, it's not my job,' replied the planter. Templer exploded. 'Well, it's true we've got some bloody bad soldiers and bad police in Malaya, but we've got some bloody bad planters as well – and you're the worst of the lot! Now get the hell out of here!'

Fighting an insurgency

Thirty-five years experience of soldiering had taught Templer some invaluable professional lessons, all of which came to the fore in Malaya. First and foremost was his understanding that without a clear political goal, there can be no military victory. Second, once he had a clear aim, he stuck to it tenaciously; he even re-read his own list of goals every morning while he shaved. Third, he believed passionately in the unity of command. Others might consider his methods high-handed, but no-one was ever in doubt as to who was in charge and what they were trying to achieve. Reluctant policemen, civil servants and soldiers were all forced to work together from joint command centres and share their information, whether they liked it or not.

Above all, Templer believed in the importance of intelligence. Not for nothing had he chased elusive guerrillas in the Judean hills and gone on to head up branches of military intelligence and SOE. 'Better a well-targeted ambush than endless jungle-bashing,' was his motto. Templer placed military intelligence officers in every static police headquarters and fused military and Special Branch intelligence into a single asset to be used to identify, track and attack the 'Communist Terrorists' (CTs) in their safe retreats. This policy allowed the army to concentrate on offensive tasks, leaving the police and local defence forces as area defence. The result was that gradually the hunters became the hunted. By the time Templer was through, every CT in Malaya was looking over his shoulder, wherever he was hiding.

This blend of politics, good intelligence and taking well-targeted operations to the enemy was highly successful. However, the key to that victory lay not just in the fighting in the jungle and the villages: Templer's real victory was that he managed to win the hearts and minds of the indigenous people. The CTs withered away, either surrendering in droves or retreating into the jungle to flee across the border.

Nevertheless, although Templer is rightly credited with turning the campaign round, this kind of war will always be a slow business. It would take another seven years before

Malaya could be declared free from communist insurgents and final victory declared.

When Templer left Malaya, the road to the airport was lined with cheering crowds of Malays and Chinese in a spontaneous demonstration of gratitude for the only commander to have crushed a full-blown communist-inspired insurgency. Templer's victory, and his methods, have much to teach us still.

Hearts and minds

From the outset the British realized that the real battle in Malaya was not for a military objective but for the popular support. Templer said that with two-thirds of the population on his side he could end the emergency in three months. He also saw that the key to winning the people over was to make them understand that the rule of law had to be observed and the government was going to win.

His secretary of defence and right-hand man, Robert Thompson, emphasized the need for government credibility. The whole policy was summed up as, 'winning the hearts and minds of the people'. In this Templer was helped by two important advantages: a well-thought-out plan and a political ace card.

The four-point plan laid the foundations for victory: to dominate the populated areas to give a feeling of security and government control; to isolate and disrupt the communist organization within those populated areas; to cut the bandits off from their food supplies; and finally to destroy the bandits by forcing them to attack the security forces on their own ground. Templer inherited this plan from the previous director of operations, but he made it work spectacularly successfully.

The key to this strategy was the 'new village programme', where nearly half a million Chinese squatters were moved out of the jungle and rehoused in new government-built protected villages. These were secured by a police force expanded to five times its 1948 level and included large areas for private cultivation. The Chinese settlers were then encouraged to become land-owners and offered citizenship. This was a masterstroke, as it effectively 'drained the sea in which the guerrillas swam' while holding out the hope of a brighter future by incorporating the Chinese into Malayan society and giving them a stake in its success.

Not everyone agreed with this forced resettlement. After one ambush in which twelve British soldiers were killed, Templer personally swooped on the village and demanded the names of those responsible. On receiving no answer, he locked down the village, restricted its food supplies and told the elders that the restrictions would stay until they gave up those responsible. These tough measures were greeted with a chorus of outrage from newspapers in Britain.

Nevertheless, Templer's grim measures worked. Gradually the locals realized that it was safer to be on Templer's side – and uncomfortable if they were not.

Templer's trump cards

These measures were backed by a revitalized intelligence service with the police and the army working together in joint operations rooms. Well-targeted special forces raids, air strikes and ambushes began to put the bandits at risk in what they thought of as 'their' jungle: but the jungle is neutral and the re-trained Commonwealth Forces learned to use it to turn the tables to hunt the CTs. Soon military casualties went down by 30 per cent and communist casualties mounted. The first trickle of surrendering bandits began to emerge from the jungle, eventually to turn into a flood.

In all this military effort Templer was helped by his political ace: nationhood. His appointment as governor-general was a civilian one and he never forgot it. His favourite theme was the new Malaya – but only when the communists were defeated. Crucially he ensured that the new Malaya would have common citizenship and independence. When he gave the Chinese the vote, the Malays called him 'pro-Chinese'; when he later brought in a modern tax structure that hit the wealthy Chinese hard, they denounced him as being 'pro-Malay'. The truth was Templer was focused on building a common new Malaysian identity and nation. His brusque military common sense told him that the days of old-style colonialism were gone and he blew away all the remnants of segregation and imperialism. He was not called the 'Tiger of Malaya' for nothing. On hearing that the Sultan of Selangor had been refused entry into the smart 'Lake Club' because of the colour of his skin, Templer reacted with fury, pointing out that there was no colour bar among his security forces risking their lives to defend the club members. The committee resigned en masse and segregation ended.

The man behind the mask

Despite his manner Gerald Templer was a man of great humanity. He genuinely cared about people: he would suddenly turn up at Malayan weddings, sing raucous army songs with the sergeants' mess and spend as much time organizing a wide-ranging and liberal programme of social legislation as on winning the fight in the jungle. Even the lowliest administrator learned to feel that Templer cared about him personally.

By the time he left Malaya in the summer of 1954 Templer had effectively broken the back of the insurgency. As he drove to the airport through cheering crowds of Malay and Chinese he genuinely could claim, 'Mission Accomplished'. His remarkable blend of intelligence, training, strategy and leadership, allied to his deft political skills and his sheer force of personality, ranks him as a commander of the highest stature. He won a crucial victory for his country and perhaps a greater one in the emergence of a new united Malaysia.

Although he may be one of Britain's least recognized generals, it is no exaggeration to say that Field Marshal Sir Gerald Templer ranks with Marlborough and Wellington as one of its most successful soldiers. The truth is that, without Templer's victory, Southeast Asia – and the post-war world – would be a very different place.

MOSHE DAYAN

1915–81

MARTIN VAN CREVELD

MOSHE DAYAN called most of his subordinates by their nicknames: 'Dado' (David Elazar), 'Arik' (Ariel Sharon), 'Talik' (Israel Tal), 'Mota' (Mordehai Gur), 'Raful' (Rafael Eytan) or 'Gandhi' (Rehavam Zeevi). The only person who dared give Dayan himself a nickname was his mother, who called him 'Musik' as a child, though later even she referred to him by his given name. From an early age he carried heavy responsibilities, yet he was a master at avoiding responsibility. He was a born leader of men, but he rarely returned the devotion those men felt for him. Affecting to be a plain-speaking soldier, he was a master of back-biting and intrigue. When circumstances demanded he could be completely ruthless, yet he enjoyed and wrote poetry like few other commanders. It was to him and others like him that the state of Israel owes its existence.

Moshe Dayan was the first child born, in 1915, to the members of the first kibbutz, Degania, on the southeastern shore of Lake Galilee. He grew up at Nahalal, a settlement not far from Haifa. It was there he went to school, though he never took his high school diploma. Most of his free time was taken up by working on the family farm. From time to time there were clashes with Arab youths from the neighbourhood; but they did not lead to any lasting resentment on Dayan's part, and he ended up by learning Arabic fairly well.

His military career started during the Arab revolt of 1936–9. The British forces were operating in the Valley of Esdraelon, trying to protect the oil pipeline that ran from Iraq to Haifa and to maintain order in general. They needed guides who knew the area, and the young Dayan qualified. One British officer in particular, Captain Orde Wingate, taught him a lot: how to make a silent approach so as to take one's enemy by surprise; how to mount an ambush; how to make prisoners talk; and, most important of all, how to kill, see one's own comrades killed, and find it within oneself to keep going.

In October 1939 he and some others were arrested by the British authorities for carrying weapons. By that time Wingate had left the country, the Arab revolt had died down and British need for Jewish allies diminished. He was sentenced to ten years in prison, later commuted to five.

Apparently it was in prison that the leadership qualities of this darkly brooding personality, who was clever, crafty and possessed of a sarcastic sense of humour, were first noticed, causing him to end up as a spokesmen for his comrades.

Young commander

He was unexpectedly released in April 1941. The British were about to invade Syria and Lebanon. Once again, they needed local Arabic-speaking guides whom they could trust. Dayan was in charge of a squad whose mission was to capture and hold a bridge on the way to Beirut. They carried out their task, but Dayan was badly wounded in the face. It was several months before he came out of hospital, minus his left eye but plus the eye-patch that was to become his trademark.

Just what he did during the next six years is not very clear. Perhaps because he was older and more experienced than most of them, the commanders of PALMACH, the pre-1948 Jewish striking force, did not want him. Apparently he worked for the intelligence branch of Hagana, PALMACH's parent organization, setting up networks of Arab informants. At one point he went to Baghdad on an intelligence mission; in 1944–5, he was in charge of arresting members of the right-wing ETZEL terrorist organization who refused to obey the Jewish community's elected leadership. Probably it was at this time that he first caught the attention of the head of the Zionist Agency and future prime minister, David Ben-Gurion, as a man who would carry out orders. To his very great credit, he did so without creating lasting enmity between himself and the men he arrested.

When Israel's War of Independence broke out in 1948, he was a major. He seems to have served as a one-man fire brigade, showing up now here, now there. In April 1948 he was instrumental in making a battalion of Druze troops that had invaded Israel from Syria change sides. Later in the same month he commanded an artillery battery that helped save his birthplace, Degania, from a Syrian offensive. May found him commanding a mechanized battalion against ETZEL, preventing them from landing an unauthorized shipment of arms at Kfar Vitkin, north of Tel Aviv. Later that month the battalion took part in fighting the Jordanians and the Egyptians. It also helped conquer the then purely Arab towns of Ramleh and Lydda, resulting in perhaps 50,000 refugees.

In June Dayan was transferred to Jerusalem, where he became city commander. By that time the fighting in the city had almost died down; two minor attacks he launched against the opposing Jordanians failed. Apparently Ben-Gurion put him in this post not because of his military abilities but mainly because of his political and diplomatic skills. The prime minister wanted somebody who could negotiate – which Dayan did, forcing the Jordanians to give up a strip of territory in the West Bank. Even more important, the prime minister needed somebody who would carry out orders and agree to leave half of Jerusalem to the enemy.

Top gun

Dayan was now recognized as a top gun, and Ben-Gurion gave him a series of appointments clearly designed to prepare him for taking over the highest post of all. First he became commander-in-chief, South, then commander-in-chief, North, and finally deputy chief of staff. In the first of these roles he used fairly brutal methods to drive out Arabs whom the state claimed should not be there, across the border into Jordan and Gaza. In the last one he had a hand in organizing raids in retaliation for the killing and wounding of Israelis by Arab 'infiltrators'. Some of the raids resulted in dozens of civilians killed. In 1955, having taken over as chief of staff, he expanded the raids into Gaza and the Sinai as well.

His military masterpiece was the Sinai campaign of October–November 1956. Israeli planning for the Sinai campaign started in October 1955, when Egypt announced a large arms deal with Czechoslovakia (in reality, the Soviet Union). From the beginning, it was hoped to involve Britain and France as well. Half of the Egyptian army was deployed west of the Suez Canal in anticipation of a Franco-British offensive, and the forces that Israel could muster were about equal to the Egyptian ones. Each side had approximately 45,000 men.

Considering how devious and unreliable Israel's allies were, designing a campaign that would make military sense while still taking account of political realities was extremely difficult. Dayan, sketching his ideas on the back of cigarette boxes, only needed about five minutes to do so. However, convincing prime minister Ben-Gurion and making the necessary preparations took longer.

By the end of October 1956 everything was ready. First, by concentrating some of his elite units on Israel's eastern border, Dayan convinced the enemy that he was about to attack Jordan. Next, a daring air force operation cut Egyptian communications in the Sinai, preventing the enemy forces from coordinating their defence.

The offensive, when it came, was launched from a totally unexpected direction. A battalion of paratroopers was dropped deep into the Sinai, confronting the enemy with a surprise from which he never recovered; next, the rest of the paratroop brigade, commanded by Lieutenant Colonel Ariel Sharon, linked up with it. First the Egyptian fortifications in central Sinai, then those in the northeastern corner of the peninsula near Rafa, were attacked and occupied. Except for one unauthorized operation which Ariel

Sharon launched into the Mitla Pass and which, though ultimately successful, carried a heavy cost in casualties, the rest was merely mopping up.

Dayan himself was 41 years old. An impetuous commander who hated meetings, he spent the campaign driving and flying all over the theatre of operations; some of the time he was out of touch with the general staff in Tel Aviv. He pushed his commanders, encouraged them, and, as happened at Abu Agheila, where the initial Israeli attack failed, fired them when he felt they deserved it. Characteristically, the end of the campaign found him at the Straits of Tiran, about as far from headquarters as he could be.

The Sinai campaign was much the smallest of the four in which Dayan played a part or which he commanded. Judging by the fact that for each of the 170 soldiers killed, three Egyptian divisions were destroyed, it was also the most successful by far. It gave the country ten years of almost uninterrupted peace. Not long after, at the beginning of 1958, he left the army. A year later he joined the Cabinet as minister of agriculture, a post he held until 1963 and in which he was not a great success.

Though he remained a member of a splinter party in the Knesset, politically he was barely active. Apparently he warned the prime minister, Levi Eshkol, and the chief of staff, Yitzhak Rabin, that their aggressive policy towards

Syria was 'madness' and would lead to war. Though married and the father of three, he entertained an endless series of mistresses.

His country's saviour: the Six Day War

Salvation came from an unexpected quarter. In April 1967 hostilities along the Israeli–Syrian border escalated. At one point, Egyptian president Gamal Abdel Nasser decided he could no longer stand aside and watch; he mobilized, re-occupied the Sinai peninsula (which had been demilitarized in 1957), and closed the Straits of Tiran to Israeli shipping. To make the crisis worse, Jordan, Syria and Iraq joined hands in declaring their intention to fight the Zionist enemy.

After three weeks of political manoeuvring, Dayan was appointed minister of defence. As he assumed the post, the entire country heaved a sigh of relief; here, at last, was a man who seemed to know what he wanted and had what it took to carry it out. Three days later, on 5 June 1967, Israel went to war.

First a devastating air strike – the plans had been prepared and rehearsed long before he took office – demolished most of the Egyptian air force on the ground. Next, three armoured divisions crashed into the Sinai, overrunning or bypassing the fortifications in their way and scattering the enemy in front of them. Eight Egyptian divisions, totalling about 100,000 men, were smashed; only a single brigade succeeded in getting away intact. Within four days, the Israelis were standing on the Suez Canal.

When the Jordanians intervened, their little air force, too, was destroyed. Israeli ground forces moved into the West Bank from three directions, north, south, and west. Those operating at the centre of the front, in and around Jerusalem, quickly surrounded the city; by so doing, they cut the West Bank in half. Though the Jordanians fought bravely, they were helpless against the Israeli air force. Within three days the entire area was occupied. Meanwhile the Israel air force also found time and aircraft with which to strike at the westernmost Iraqi air base from which attacks had been launched on Tel Aviv.

Throughout the four days these operations lasted, the Syrians, from their fortified positions on the Golan Heights, rained down artillery shells on the Israeli settlements below. The general staff, as well as members of the Cabinet, demanded that Dayan authorize an attack on the Heights; fearing Soviet intervention, he refused repeatedly. Once Egypt agreed to a ceasefire during the night of 8 June, however, he changed his mind, ordering the assault without bothering to inform the prime minister. Later, he was often blamed for this.

As in 1956, he took the most important decisions as to which enemy to fight, where to fight him, and, perhaps most important of all, where to stop. As in 1956, he spent as much time as he could away from headquarters. He toured the fronts and gave directives on the spot; one result of this was that, when Israeli aircraft mistakenly attacked the USS *Liberty*, he could not be reached. Some of the decisions he made, such as leaving the Temple Mount under the control of the Muslim Waqf (religious endowment) were to have consequences that last to the present day.

Towards the Day of Judgment

In 1967–73, Dayan presided over a great expansion of the Israel Defense Force (IDF), whose order of battle grew by about 60 per cent. Many new weapons were purchased, and qualitative improvements introduced. The result was to turn a force which, until then, had been very well motivated and trained but poor and technologically somewhat backward, into a regional juggernaut.

The years 1967–70 in particular were anything but quiet. A military administration had to be put in place in the occupied territories. There were numerous skirmishes with Jordanian and Syrian forces as well as, increasingly, Palestinian guerrillas operating first from Jordan and then from Lebanon.

The largest and most dangerous clashes took place along the Suez Canal from March 1969 to August 1970.

Relatively small Israeli forces, hunkering in their bunkers, confronted an enormous Egyptian force equipped with as many as 1,000 artillery pieces. To restore the balance, Dayan ordered the Israeli air force into action. This in turn caused the Soviets to intervene – they already had 'advisers' at every level of the Egyptian army. In April 1970 several Soviet fighters were shot down. For a time, it looked as if the mighty Soviet Union would be drawn into a war against tiny Israel.

The pressure on Dayan during those years was enormous. To make things worse, in 1968 he was nearly buried alive during an archaeological dig, receiving injuries from which he never quite recovered. Once a ceasefire was agreed on along the Suez Canal in August 1970, things became somewhat easier. Too easy perhaps; when a quarter of a million Arab troops with 5,000 tanks between them attacked in October 1973, they found Israel unprepared.

The Yom Kippur War

Dayan, in fact, had suspected there might be a war. In the spring he repeatedly warned the General Staff to prepare; however, later in the summer, when no Arab attack materialized, he calmed down. During the last few days before the Egyptian–Syrian attack he held frantic consultations with the prime minister as well as his main military advisers, but in the end no decision to mobilize was made. As a result, on 6 October 1973, Yom Kippur or the Jewish Day of Judgment, the Egyptians and the Syrians took Israel by surprise just as Israel had surprised them six years earlier.

Early in the war, it was Dayan who had to make the most critical decisions. Which front, the Egyptian or the Syrian one, to give priority; whether to try to hold ground or retreat to a line further in the rear; whether to use the air force to attack the enemy's anti-aircraft systems, as planned, or sacrifice it in a desperate attempt to stop the enemy's advance; and so on. As was his habit, he preferred to make these and similar decisions not in the comfort of general headquarters but on the basis of visits to the front. Later Ariel Sharon, who commanded a division in the south, recalled that Dayan had been the only senior commander to visit him.

Seen in retrospect, some of the decisions he made during those days may not have been correct. It seems that there was a moment on Sunday 7 October or Monday 8 October, when things looked so bad that Dayan suggested threatening Syria with Israel's nuclear weapons. Whether the threat was made, and what effect it had, remains shrouded in secrecy to the present day.

From Wednesday 10 October the fronts stabilized. Dayan offered Prime Minister Golda Meir his resignation, but was told to stay at his post. While he went on visiting the fronts, others, including above all chief of staff General David Elazar, now made the most important decisions. When the war ended on 24 October, the Syrians had been thrown back, about a third of the Egyptian army was surrounded, and Israeli troops stood west of the Suez Canal, just over 60 miles from Cairo.

Final years

When the commission set up to investigate the war published its findings, it did not condemn Dayan – the supreme example of his ability to evade responsibility. Public opinion nonetheless forced him to resign. In June 1974, though still a member of the Knesset, he returned home and spent the next eighteen months writing his memoirs, a fascinating account by a difficult and complex man.

In May 1977 Israel's new prime minister Menahem Begin asked Dayan to serve as foreign minister. In this capacity, he played a critical part in negotiating peace with Egypt at Camp David (September 1978). Soon thereafter, realizing that Begin had no intention of continuing the peace process so a solution might be found that would rid Israel of its 'hump' – the occupied West Bank – he resigned. By that time he was a sick man, suffering from colon cancer. He tried to warn Begin against launching a war in Lebanon, but did not live to see the failure of his efforts. On 16

October 1981 he died. He was buried in Nahalal, close to his parents and other relatives. Proud to the last, he had asked that no eulogies be held on his grave, and his wish was respected.

VO NGUYEN GIAP

1911–

GEOFFREY PERRET

GIAP STUDIED THE ACHIEVEMENTS of such great commanders as Napoleon and Mao, and applied the lessons to the anti-colonial struggle in Vietnam. Taking advantage of the huge loyalty and enthusiasm of the Vietnamese peasantry whom he adulated, Giap built up an army from nothing, and with it he defeated two great powers, France and the United States.

Vo Nguyen Giap was born in August 1911 in An Xa, a small coastal town in the province of Quang Binh, in the narrow waist of Vietnam. His father was a minor civil servant and land-owner.

The youthful Giap's course in life was shaped largely by the French occupation. For much of the nineteenth century France had struggled to bring Indochina into its empire. By 1885 France controlled Indochina, which consisted mainly of French protectorates, plus the colony of Cochinchina. Although they justified their rule as '*une mission civilatrice*', spreading the values of the Enlightenment to inferior peoples, the true pillars of French rule were systematic torture, harsh prison sentences and Madame Guillotine. Seeking a cowed population, what they got instead was resistance movements.

Given the region's history, warrior motifs had ineluctably worked their way into the names of people and places. Giap's family name of Vo, for example, can be read as 'military force', while Giap means both 'first' and 'intact armour'. Some Vietnamese even claim that his name in its entirety means 'commander-in-chief'.

The making of a revolutionary

More important than the symbolic references contained within his names were the fervently anti-French sentiments of Giap's father. He recited patriotic poems to his son, taught him songs of sacrifice and courage and talked about the many wars of Vietnam, fought mainly against the Chinese. Throughout his life Giap would remain wary of the great neighbour to the north.

Even so, the collapse of imperial China in the same year that Giap was born brought more than thirty years of civil strife that wracked every major Chinese province, city and large town. China's internal struggles inflamed patriotic and nationalistic movements of every variety, from anarchism to Marxism to crypto-fascism. Sparks cast out from these struggles had incendiary effects in neighbouring Vietnam.

As childhood gave way to adolescence the young Giap was not so much pushed into anti-French politics as rushing to embrace it. Exceptionally intelligent and hard-working, at the age of 14 he won admission to one of the most prestigious secondary schools in the country, the Lycée Quoc Hoc in Hué. The school's graduates included Ho Chi Minh and Ngo Dinh Diem.

One day a fellow student handed him a proscribed pamphlet entitled 'French Colonialism on Trial'. Giap climbed into a tree to read it undisturbed. He descended filled with rage, but thrilled to his patriotic core. The author was Ho Chi Minh, although Giap did not know that then as it had been published under one of Ho's many aliases.

In 1930 the French arrested seven teachers at Giap's school and a number of students were rounded up, including Giap. He was sentenced to two years' hard labour. When he left prison, however, he was offered a second chance. A senior French official encouraged him to apply to the Lycée Albert Sarraut in Hanoi, the Vietnamese equivalent of Eton. The French routinely attempted to co-opt the future leaders and opinion-shapers among their colonial subjects while they were still at a formative stage, with careers to make.

Vive l'Empereur!

In 1934, having gained his baccalaureate, Giap taught history and French at a *lycée* while studying law at the University of Hanoi. By now, he was a dedicated communist, although as with most Vietnamese communists, his was a communism that was grafted on to nationalistic rootstock.

Meanwhile he was cultivating intellectual interests that had little or no connection with the legal profession. At the start of the history course he taught, he informed his students, 'I'm going to tell you about two things – the French Revolution and Napoleon.' Throughout his long life, every one of the emperor's battles and campaigns fascinated Giap and inspired him. After all, Napoleon was a great revolutionary as well as being a great soldier.

In 1938, as the number one student in his law school class, Giap was offered a full scholarship to attend any of the *grandes écoles* or universities of France. There he would be free to study whatever he desired. He rejected the offer as a matter of course. By this time, Giap knew just what his life's work would be: to get the French out of Indochina. He turned his talents to waging a propaganda offensive. With a group of like-minded friends Giap bought a failing newspaper and devoted its pages to tales of French perfidy and Vietnamese courage. Although his prose style was larded with the mind-numbing jargon of Marxism-Leninism he saw himself as a writer by vocation and introduced himself to people as a journalist.

Giap also travelled to China, where he finally met the author of 'French Colonialism on Trial'. Ho Chi Minh – meaning 'the enlightened one' – was by this time the best-known figure in the growing resistance to French rule. Ho and Giap became so close they might as well have been father and son. During this Chinese journey Giap also came into contact with one of the most famous military formations of all time, the Eighth Route Army, the creation of Mao Zedong and a professional soldier, Zhu De. Following a failed attempt to engineer a peasant uprising in his home province of Hunan in 1926, Mao had concluded that for the Chinese Communist Party to survive it must build its own army. Zhu was ready to help, but where would they find troops?

Peasant power

Mao had the answer: the despised peasants of China constituted the most powerful revolutionary force imaginable. Organize them, mobilize them, inspire them, and these peasants, the truly wretched of the Earth, could defeat anyone. They had little to lose but their lives, but under most Chinese governments, those lives were hardly worth living anyway. Mao taught his soldiers that their mission in life was to protect the peasantry against both the rapacious Nationalist forces of Jiang Jieshi (Chiang Kai-shek) and the invading armies of imperial Japan.

Mao's soldiers were taught to take nothing from the people, not even a needle and thread. In learning to respect others, they began to respect themselves. There will always be men who are willing die for an idea, and Mao gave them one. He also gave them a military education that any illiterate peasant could understand:

> Enemy advances, we retreat.
> Enemy halts, we harass.
> Enemy tires, we attack.
> Enemy retreats, we pursue.

Giap was also influenced by Napoleon's approach to battle. Strategically, this meant a belief that only the offensive is decisive in war. Tactically, it meant, as Napoleon expressed it, 'Get stuck in and see what follows.' It was risky, opportunistic and only someone able to read the fog of war like a map could ever hope to master it.

Meanwhile, in 1939 Giap married and the next year his wife Quang Thai had a child, a baby girl. With the fall of France in June 1940 the French hold on Indochina became increasingly tenuous. They began to mount large-scale roundups of known and suspected critics of their rule. They seized Quang Thai but Giap escaped his pursuers. Quang Thai's sister was seized, tortured and shot. Quang Thai herself was tortured in prison and perished there.

In 1941 the Japanese arrived in Vietnam but Giap did not believe they would be there long. The Axis countries – Germany, Italy and Japan – could never hope to prevail against an alliance based on the United States, the British Empire and the Soviet Union. From a rudimentary camp only a few miles into China, Giap prepared himself for the coming struggle. Yet he was taught how to organize and wage a guerrilla struggle by veterans of the Eighth Route Army. He also studied Mao's extensive writings on protracted war, including such classic texts as *A Single Spark Can Start a Prairie Fire*.

The traditional Marxist theory of armed struggle called for organizing the industrial workers. China had virtually no industrial workers. Nor had Vietnam. However, Mao had already demonstrated that in a peasant-based society it was the peasants who held the keys to power. It was impossible, though, for Giap to acknowledge his debt to Mao Zedong. Instead, he claimed that the doctrine of protracted war and the crucial role of the peasant were ideas that originated in Vietnam, with an obscure revolutionary named Bac Ho, whom very few have ever heard of.

Unable to organize in areas under French or Japanese control, Giap established himself in remote and mountainous border regions populated mainly by non-Vietnamese people who spoke their own tribal languages. Giap learned two of these languages, won their trust and set up military training camps on both sides of the border with China. Meanwhile Ho Chi Minh was creating a revolutionary political arm, the Viet Minh. Ho believed it would be possible to negotiate the French out of Vietnam. There did not have to be a war.

Armed propaganda brigades

The compromise between Ho's views and Giap's was the formation of 'Armed Propaganda Brigades'. These would wage a propaganda offensive and, when the opportunity arose, mount military attacks on isolated French police posts. Giap chose thirty-one men and three women to form the first brigade. It boasted a single light machine-gun, seventeen rifles, two handguns and some ancient shotguns and pistols.

By December 1944 Giap was ready to mount his first military assault, striking a regional French police headquarters on Christmas Eve, when the French were likely to be inebriated and easy to surprise. The headquarters was captured and the French commander perished in its defence. An almost laughably small force of thirty-four lightly armed volunteers thus became the foundation-stone of what eventually became the People's Army of Vietnam (PAVN).

By the time of Japan's surrender in August 1945 much of Vietnam, especially in the north, was firmly under Viet Minh control, allowing Ho to proclaim the formation of a national government in Hanoi. Giap could now expand the work of his armed propaganda brigades across the countryside, creating guerrilla fighters and supporters everywhere. As he later expressed it, 'Each man was a soldier. Each village or hamlet was a fortress.'

In 1946 the French seized Hanoi. Ho and the Viet Minh took to the mountains and jungles and an ever-growing army of guerrillas bled and stretched the French. Giap was also winning the most important side of the unfolding war, the struggle for moral supremacy. As that balance shifted, people across France began to doubt both the justice and the ultimate winnability of the war.

Red River campaign

The army he had crafted possessed what commanders down the ages have prized most – an indomitable fighting spirit, sustained by high morale. In January 1951 Giap launched a frontal assault to take Hanoi back from the French.

Impatient and inexperienced, he accepted the advice of Chinese military advisers and mounted a human-wave attack in broad daylight, thrusting 20,000 soldiers into a killing-ground that the French had prepared. In a matter of days Giap suffered at least 10,000 men killed, wounded or captured. He had also failed to recognize that during the battle the French commander had made a serious mistake, one that could have put victory within Giap's reach.

Refusing to admit defeat, two months later he tried again, with a three-division attack. This time he called it off after losing 3,000 troops. Undeterred, Giap mounted yet another major attack, albeit from a different direction, and lost another 10,000 soldiers to death, wounds and captivity. When he did the same again, for a fourth time, some 9,000 PAVN soldiers perished.

Morale crumbled and desertions rose. Had it not been for Ho Chi Minh, Giap would have been removed from his command. By this time, however, he had learned how to move his army, how to conceal it, how to supply it and how to use it. Without his failed Red River campaign of 1951–2, Giap would never have been able to pull off his masterstroke, the capture of Dien Bien Phu. He had also learned at last the wisdom of the military adage that says, 'Amateurs talk tactics, professionals talk logistics'.

For France, the Indochinese struggle had become a political disaster at home and abroad. They were almost desperate to hand the bulk of the fighting over to a Vietnamese army that was French-trained and French-equipped.

That would allow Paris to withdraw most of its troops but still dominate Vietnam politically.

In 1953, the new French commander in Indochina, Henri Navarre, proposed to defend northern Laos. He would spend 1954 building up a cordon of mutually supporting bases. Then, in 1955, he would thrust into the areas that Giap controlled and destroy the PAVN's main force units. Like other French generals before him and American generals after him, Navarre liked to think that he held both the tactical and strategic initiative. In a conventional war, this might have been true. In the kind of war that Giap was waging, it was he, not Navarre, who held the initiative, always.

Dien Bien Phu

The key to Navarre's strategy was the large base he was building at Dien Bien Phu, close to the northern Vietnamese border with Laos. With insouciance verging on contempt, Navarre and other senior officers convinced themselves that they did not have to occupy the high ground.

Mao meanwhile offered Giap twenty-four of the 105mm howitzers captured by Chinese troops from American units in Korea. He also offered hundreds of tons of 105mm ammunition. But how would Giap get these heavy pieces up on those jungle-clad hills? How would he move all that ammunition? Giap appealed to the peasants and nearly 1 million volunteered to serve as porters.

The Viet Minh meanwhile were mounting so many ambushes that the French tried to avoid using the roads. Giap was turning Dien Bien Phu into a virtual island, one that could be reached only by air.

When the fight opened on 13 March 1954, Giap had 49,000 soldiers to pit against Navarre's 11,000 at Dien Bien Phu, plus 2,000 French reinforcements who arrived in the course of the fight. The Chinese advisers at Giap's headquarters, who included Zhu De, urged him to mount human-wave attacks, but he had learned the cost of their advice during the Red River campaign. Despite angry lectures from Chinese generals, he held fast.

Giap knew his men would fight to the death, but he wanted them to fight to victory. He would take Dien Bien Phu by classic siege tactics, including digging trenches ever closer to the objective. He had millions of rounds of small-arms ammunition, nearly 200 artillery pieces, 60,000 rounds for his artillery, and a small mountain of rice. He was also going to need 4,000 tons of supplies to reach his army every day, much of it only after being carried for hundreds of miles on bicycles.

Giap concentrated overwhelming forces at each of the strong-points that ringed Dien Bien Phu, seized one then moved on to the next. He was also sending men to make attacks on French installations deep in the enemy's rear, striking airfields and barracks, pinning the French down not only at Dien Bien Phu but across Indochina.

Meanwhile the 105s fired day and night. The French wounded were flown out and reinforcements were flown in – some even parachuted in – until the Chinese provided anti-aircraft guns. After that there were few flights in either direction. Casualties among the defenders were comparatively light but their food, water and medical supplies were running out, just as a conference on the future of Indochina was about to be convened in Geneva. Giap attacked with all that he had. The French broke first. Dien Bien Phu surrendered on 7 May 1954, the day before the Geneva Conference convened. Giap had incurred around 23,000 casualties; the French, approximately 8,000.

The aftermath

After Giap's triumph at Dien Bien Phu the French quickly departed from Indochina, but the victory did not yet lead to the unification of Vietnam and, within the Viet Minh elite Giap had accumulated powerful rivals and critics.

Communist governments are always fearful of 'Bonapartism' – the glamorous, popular general who stages a coup. Giap was the kind of great captain that the grey politicians needed but sought to limit once victory was won. His preference for the Soviets over the Chinese also carried a price.

Giap was consulted on future operations, but he was no longer commander-in-chief. He was, for example, extremely dubious about the Tet Offensive against the Americans in 1968. The plan was predicated on a popular uprising in support of the opening round of attacks, something Giap doubted would happen. All the same, once the decision was made, he loyally did all he could to make the offensive a success.

He fooled the American commander General William C. Westmoreland into pushing troops and air assets into a major fight at Khe Sanh, but the offensive was aimed at the cities and major towns of South Vietnam; Khe Sanh was just a diversion to pull American troops away from heavily populated areas. The base depended on a small river that flowed down from the north for its water supplies. Giap could have made the base untenable by poisoning the river. The fact that he did not do so revealed his true intentions, but Westmoreland never realized that.

In his book *People's War, People's Army* (1962) Giap had explained his strategy for victory: 'Military action was necessary, but propaganda was much more important.' Westmoreland never understood that, nor did Navarre, but Tet proved Giap's point, for a second time. In 1975 the army and the guerrilla forces that Giap had created destroyed the American-backed government in Saigon. Vietnam was finally united, and in the manner that Giap had preached from the beginning. His was the strategy, ultimately, of the greatest of all military thinkers – Sun Tzu: defeat your enemy in his mind first, the rest will follow.

In the end, Giap could claim to have defeated eight French generals, of whom Navarre was only the last. After that he defeated four American generals. And now, in his 100th year, he is honoured around the world. In 1966, as an American army deployed across South Vietnam, *Time* magazine warned what they were up against. Giap, it declared, was 'the Red Napoleon'. For him, there could be no greater accolade.

FURTHER READING

HELMUTH VON MOLTKE

Lieutenant Colonel F. E. Whitton, *Moltke* (Constable, London, 1921).

Peter Paret (ed.), *Makers of Modern Strategy* (Oxford University Press, Oxford, 1986).

Michael Howard, *The Franco-Prussian War* (Hart-Davis, London, 1962).

GARNET WOLSELEY

Brian Bond (ed.), *Victorian Military Campaigns* (Hutchinson, London, 1967).

Halik Kochanski, *Sir Garnet Wolseley: Victorian Hero* (Hambledon Continuum, London, 1999).

Joseph Lehmann, *All Sir Garnet: A Life of Field Marshal Lord Wolseley* (Jonathan Cape, London, 1964).

ERICH LUDENDORFF

John Lee, *The Warlords: Hindenburg and Ludendorff* (Weidenfeld & Nicolson, London, 2005).

Correlli Barnett, *The Swordbearers* (Cassell, London, 2000).

Peter Paret (ed.), *Makers of Modern Strategy* (Oxford University Press, Oxford, 1986).

FERDINAND FOCH

Ferdinand Foch, *The Principles of War* (Kessinger Publishing, Whitefish, MT, 2007).

Michael S. Neiberg, *Foch: Supreme Allied Commander in the Great War* (Brassey's US, Washington, DC, 2004).

Basil Liddell Hart, *Foch: Man of Orleans* (Kessinger Publishing, Whitefish, MT, 2008).

PHILIPPE PÉTAIN

C. Williams, *Pétain* (Little, Brown, London, 2005).

G. Pedroncini, *Pétain, le soldat et la gloire, 1856–1918* (Perrin, Paris, 1989).

S. Ryan, *Pétain the Soldier* (A.S. Barnes & Co., New York, 1969).

EDMUND ALLENBY

M. Hughes, *Allenby and British Strategy in the Middle East, 1917–1919* (Routledge, London, 1999).

L. James, *Imperial Warrior: The Life and Times of Field-Marshal Viscount Allenby, 1861–1936* (Weidenfeld & Nicolson, London, 1993).

JOHN PERSHING

Frank E. Vandiver, *Black Jack: The Life and Times of John J. Pershing* (2 vols, Texas A & M University Press, College Station, Texas and London, 1977).

John J. Pershing, *My Experiences in the World War* (2 vols, Frederick A. Stokes, New York, 1931).

Donald Smythe, *Pershing: General of the Armies* (Indiana University Press, Bloomington, 1986).

Thomas Fleming, 'Iron General', *Military History Quarterly*, vol. 7, no. 2 (Winter 1994).

KEMAL ATATÜRK

Lord Kinross, *Atatürk: The Rebirth of a Nation* (Weidenfeld & Nicolson, London, 1964).

A. L. Macfie, *Atatürk* (Longman, Harlow, 1994).

Andrew Mango, *Atatürk* (John Murray, London, 1999).

BASIL LIDDELL HART

Alex Danchev, *Alchemist of War: The Life of Basil Liddell Hart* (Weidenfeld & Nicolson, London, 1998).

Sir Basil Liddell Hart, *The Memoirs of Captain Liddell Hart* (Cassell, London, 1965).

Sir Basil Liddell Hart, *Great Captains Unveiled* (Presidio Press, Novato, CA, 1990).

John J. Mearsheimer, *Liddell Hart and the Weight of History* (Cornell University Press, Ithaca, NY, 1993).

CARL GUSTAF MANNERHEIM

H. M. Tillotson, *Finland at Peace and War* (Michael Russell, London, 1993).

J. E. O. Screen, *Mannerheim: the Years of Preparation* (Hurst, London, 1970).

Oliver Warner, *Marshal Mannerheim and the Finns* (Weidenfeld & Nicolson, London, 1967).

Stig Jägerskiöld, *Mannerheim: Marshal of Finland* (Hurst, London, 1986).

GERD VON RUNDSTEDT

Charles Messenger, *The Last Prussian: A Biography of Field Marshal Gerd von Rundstedt 1875–1953* (Elsevier, London, 1991).

Earl F. Ziemke, 'Field Marshal Gerd von Rundstedt', in Correlli Barnett (ed.), *Hitler's Generals* (Grove Press, London, 1989), pp. 175–207.

ERICH VON MANSTEIN

Enrico Syring and Ronald Smelser (eds), *Die Militarelite des Dritten Reiches*, 'Erich von Manstein – Das Operative Genie' (Ullstein, Berlin, 1995), pp. 325–48.

Oliver von Wrochem, *Erich von Manstein. Vernichtungskrieg und Geschichtspolitik* (Schöningh, Paderborn, 2006).

HEINZ GUDERIAN

Heinz Guderian, *Panzer Leader* (Michael Joseph, London, 1952).

Kenneth Macksey, *Guderian* (Greenhill, London, 1992).

ERWIN ROMMEL

Charles Douglas-Home, 'Field Marshal Erwin Rommel', in Michael Carver (ed), *The War Lords* (Weidenfeld & Nicolson, London, 1976).

Rick Atkinson, *An Army at Dawn: The War in North Africa 1942–1943* (Little, Brown, London, 2003).

David Fraser, *Knight's Cross: The Life of Field Marshal Erwin Rommel* (HarperCollins, London, 1993).

Samuel Mitcham, *The Desert Fox in Normandy: Rommel's Defence of Fortress Europe* (Cooper Square Press, London, 2001).

BERNARD MONTGOMERY

Bernard Montgomery, *The Memoirs of Field-Marshal the Viscount Montgomery of Alamein* (William Collins, Glasgow, 1958).

Nigel Hamilton, *Monty, The Life of Montgomery of Alamein: 1887–1942* (Hamish Hamilton, London, 1981).

Nigel Hamilton, *Monty, The Life of Montgomery of Alamein: 1942–44* (Hamish Hamilton, London, 1983).

Nigel Hamilton, *Monty, The Life of Montgomery of Alamein: 1944–76* (Hamish Hamilton, London, 1986).

GEORGI ZHUKOV

Robert Service, *Stalin* (Macmillan, London, 2004).

G. K. Zhukov, *The Memoirs of Marshal Zhukov* (Jonathan Cape, London, 1971).

Andrew Nagorski, *The Greatest Battle: The Battle of Moscow 1941–2* (Aurum Press, London, 2007).

Chris Bellamy, *Absolute War: Soviet Russia in the Second World War* (Macmillan, London, 2007).

William J. Spahr, *Zhukov: The Rise and Fall of a Great Captain* (Presidio Press, Novato, CA, 1993).

Antony Beevor, *Stalingrad* (Viking, London, 1999).

David M. Glantz, *Zhukov's Greatest Defeat: The Red Army's Epic Disaster in Operation Mars* (University Press of Kansan, Lawrence, KS, 1999).

IVAN KONEV

Ivan Konev, *Year of Victory* (Progress Publishers, Moscow, 1969).

Ivan Konev, 'The Korsun–Shevchenkovsky Pocket', in *Battles Hitler Lost* (Richardson and Steirman, New York, 1986), pp. 111–26.

Evan Mawdsley, *Thunder in the East: The Nazi–Soviet War 1941–1945* (Hodder Arnold, London, 2005).

Harold Shukman (ed.), *Stalin's Generals* (Weidenfeld & Nicolson, London, 1993).

David Glantz, Jonathan House, *When Titans Clashed: How*

the Red Army Stopped Hitler (University Press of Kansas, Lawrence, KS, 1995).

DWIGHT D. EISENHOWER

Carlo D'Este, *Eisenhower: Allied Supreme Commander* (Weidenfeld & Nicolson, London, 2002).

Geoffrey Perret, *Eisenhower* (Random House, New York, 1999).

GEORGE S. PATTON

Martin Blumenson, *Patton: The Man Behind the Legend* (Jonathan Cape, London, 1986).

Carlo D'Este, *A Genius for War: A Life of George S. Patton* (HarperCollins, London, 1995).

Trevor Royle, *Patton: Old Blood and Guts* (Weidenfeld & Nicolson, London, 2005).

TOMOYUKI YAMASHITA

J. D. Potter, *A Soldier Must Hang* (Muller, London, 1963).

Akashi Yoji, 'General Yamashita Tomoyuki: Commander of the 25th Army', in Brian P. Farrell and Sandy Hunter (eds), *Sixty Years On: The Fall of Singapore Revisited* (Times Academic Press, Singapore, 2002).

DOUGLAS MACARTHUR

D. Clayton James, *The Years of MacArthur* (2 vols, Houghton Mifflin, Boston, 1970 and 1975).

Geoffrey Perret, *Old Soldiers Never Die: The Life of Douglas MacArthur* (Random House, New York, 1996).

WILLIAM SLIM

Field Marshal Viscount Slim, *Defeat into Victory: Battling Japan in Burma and India 1942–45* (Cooper Square Press, London, 2000).

Robert Lyman, *Slim: Master of War, Burma and the Birth of Modern Warfare* (Constable, London, 2005).

Ronald Lewin, *Slim: The Standardbearer, A Biography of Field-Marshal The Viscount Slim* (Leo Cooper, London, 1976).

GERALD TEMPLER

John Cloake, *Templer, Tiger of Malaya: The Life of Field Marshal Sir Gerald Templer* (Harrap, London, 1985).

Robin Neillands, *A Fighting Retreat: The British Empire 1947–97* (Hodder & Stoughton, London, 1997).

Brian Lapping, *End of Empire* (St Martin's Press, New York, 1985).

Anthony Heathcote, *The British Field Marshals 1736–1997* (Pen & Sword Books Ltd, Barnsley, 1999).

Kumar Ramakrishna, *Emergency Propaganda: The Winning of Malayan Hearts and Minds 1948–1958* (Curzon Press, Richmond, 2002).

MOSHE DAYAN

Moshe Dayan, *Story of My Life* (Weidenfeld & Nicolson, London, 1976).

Martin van Creveld, *Moshe Dayan* (Cassell, London, 2002).

VO NGUYEN GIAP

Cecil B. Currey, *Victory at Any Cost: The Genius of Vietnam's Vo Nguyen Giap* (Potomac Books, Washington, DC, 1997).

Robert J. O'Neill, *General Giap: Politician and Strategist* (Praeger, New York, 1969).

Vo Nguyen Giap, *People's War, People's Army* (Praeger, New York, 1962).

INDEX